The Digital Ice Age

by Rolf A. F. Witzsche

Contents

On the Ice Age and Climate Change
and the book

The Digital Ice Age

Book 2 of the series: Ice Age of the Dimmer Sun in 30 Years

The term, 'Digital', refers to on-off conditions. Computer electronics is built on it. Cosmic 'electronics' reflect the principle on a larger scale. The 'Primer Fields' that focus interstellar plasma onto our Sun are dynamically formed and are subject to on-off conditions. Ice Ages result in the off periods.

Without the Primer Fields, the Earth would be in an Ice Age environment. We, ourselves, would likely not exist then. The fields are created by magnetic effects of flowing plasma acting on itself. The phenomenon has been explored and replicated in laboratory experiments. The characteristics are presented in this book.

A threshold level in plasma density must be exceeded for the fields to form. In times when the fields cannot form, or can no longer be maintained, the Sun is powered at a lower level. During the inactive state the solar activity is reduced to a type of cosmic default level with 70% less radiated energy. This has been the case for 85% of the last 2 million years.

Ice Age Glaciation is the normal state on Earth, except for the brief 'active' intervals of interglacial periods and the Dansgaard Oeschger oscillations during the Ice Ages.

At the present rate of diminishment, the solar activity phase-shift threshold may be crossed in 30 years, or in the 2050s, most likely. With the primer system gone inactive, the climate on Earth will get 40 times colder than the Little Ice Age in the 1600s had been. Ice core evidence promises that. Without the needed preparations for human living in such an environment, 99% of humanity would die of starvation, both by the cold and by CO_2 depletion as more CO_2 becomes dissolved into the sea.

With the fields being critical for our very existence, the exploration of it is likewise critical.

In the Little Ice Age, between 10% and up to 30% of the populations in Europe had perished by starvation. The last Big Ice Age was evidently vastly harsher. Only 1-10 million people emerged from it alive. That's all we had after 2 million years of development. We want to do far better this time around; and we can, with large-scale technological infrastructures for our food supply. But will we create them? Will we get the job done in the 30 years that we still have left before the Ice Age starts anew? Will we even consider it? And how certain are we that the phase shift to the next glaciation period will begin, as the evidence suggests, in the 2050s? We have no slack on this front. Should we fail us on this absolute front, we would be committing suicide.

Numerous fields of evidence tell us that the next Ice Age is near. That's where the truth begins. Most of the evidence was discovered in the 1990s and thereafter. Some evidence is measured in ice cores; some is measured in space, by satellites. Some measurements are also made on the ground in terms of measurements of the Earth's magnetic-pole drift observed in northern Canada. All of this is seen combined with high-energy physics experiments at a leading national laboratory, and is also explored in the small in static experiments.

So, what will the answer be? Will we move with the evidence? Or will we lay ourselves down to die by default?

It takes an independent researcher to brake the taboos that have kept mainstream cosmology imprisoned, increasingly, during the past century, even while what is regarded as taboo is known to be wrong.

The Illustrated Science series is intended to open the scene beyond the threshold of accepted taboos, to where the actual physical evidence speaks for itself.

The scope of the existential challenge that the Ice Age brings with it, takes astrophysics out of the academic domain and places it into the foreground as one of the most-critical issues of our time. The big Climate Change events that have already worldwide effects are mere fringe effects in the flow of the ever-changing cosmic dynamics. The big effect, when the Ice

Age begins anew, promises to be caused by a dimmer and colder Sun. The loss of 70% of the Sun's radiated energy defines our climate future that begins in the near term.

Sure, we can live with all that by creating new platforms for agriculture that are able to operate under Ice Age conditions. But will we do it? The task is enormous. Or will we fail ourselves on this front? We have no reason to allow us to fail. We have the materials and energy resources on hand to accomplish everything that is required for us to continue to live in an Ice Age World. But will we do it? The big question that never goes away, therefore, is; will we develop our inner resources as human beings sufficiently to get the job done, and to get it done in time? Or will we do nothing, ignore the challenge, and condemn our children and one-another to an agonizing death by starvation? That's the choice.

Towards meeting the inner challenge, I have created the epic series of novels, The Lodging for the Rose. And further, towards meeting the science challenge, I have produced numerous research books and several dozen exploration videos that the Illustrated Science series is modeled after. The work is the result of a quarter century of research, for which numerous elements of evidence in related fields came to light during the timeframe of my research.

It is my hope that the work that went into all of these projects will help in some degree - for humanity that we are all a part of - to write itself a ticket to have a future.

High-resolution color images, of the images in this book, can be obtained at www.iceagetheatre.ca

The digital Ice Age

The two long climate cycles that overlap

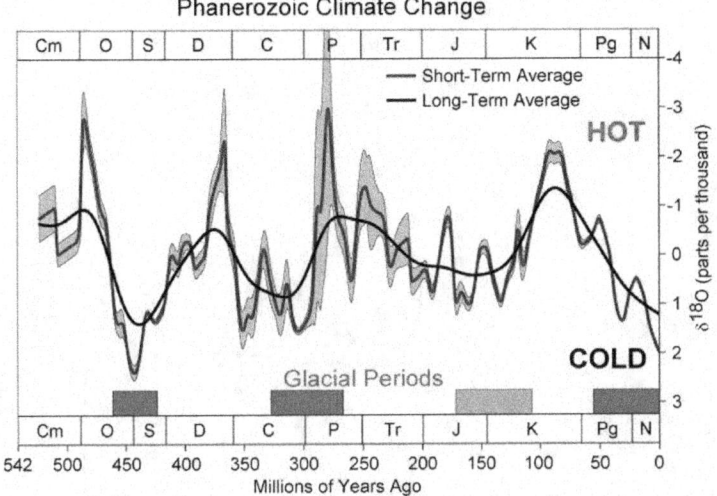

Phanerozoic Climate Change

Major glaciation periods have occurred 4 times in the last half-billion years that we have records of. Some were so severe that major extinctions occurred as a consequence. We are presently in such a period that is extremely severe. The two long climate cycles that overlap and determine the strength of our solar system, are both approaching their minimal point together.

In the last 500,000 years of the resulting glaciation epoch

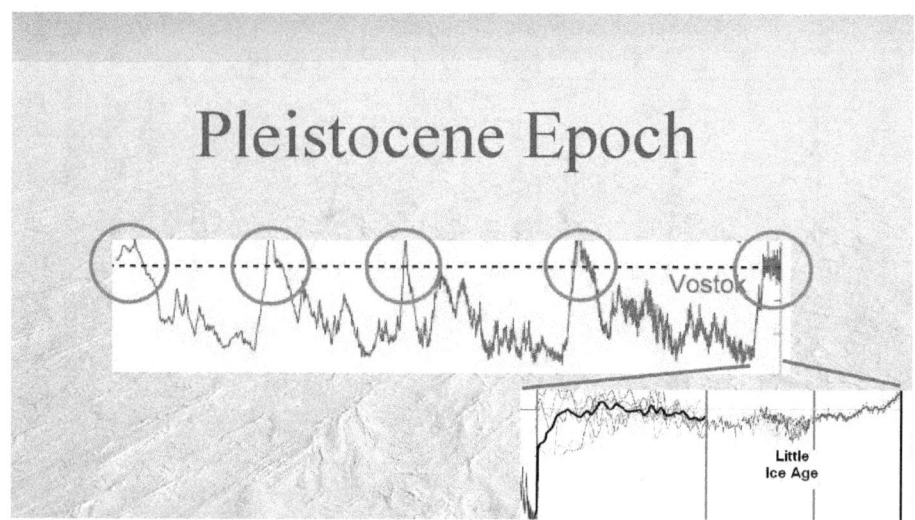

In the last 500,000 years of the resulting glaciation epoch, glaciation conditions occurred for 85% of the time, interspersed with brief interglacial warm periods, like the one we presently enjoy, which we erroneously regard as normal, but which has run its course and is now ending.
This means that the current warm period is a climate anomaly that is actually rather fragile. The Ice Age conditions are the normal state on Earth.

The difference between the two climate states

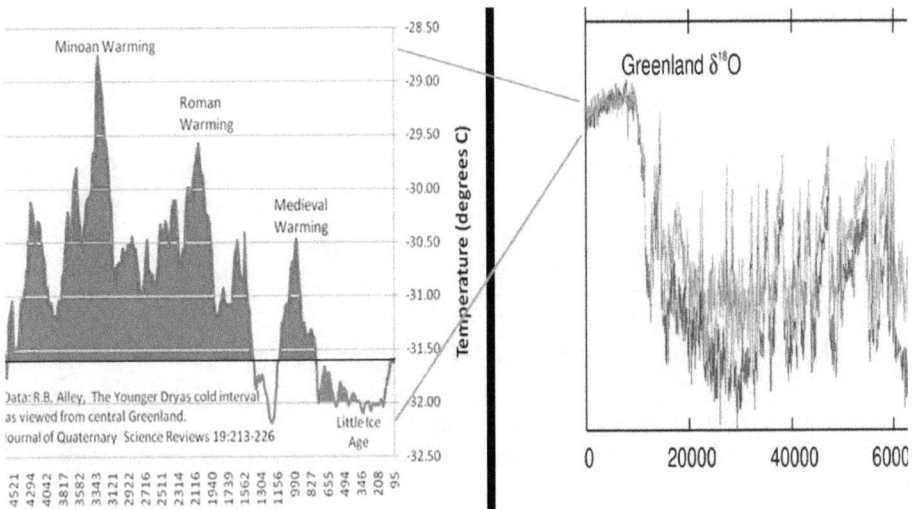

The difference between the two climate states is so enormously large, and occurs with such a swift transition between them, that it can only be rationally understood as the result of solar on-off transition. That's what the ice core records indicate.

Produced from two different drilling sites

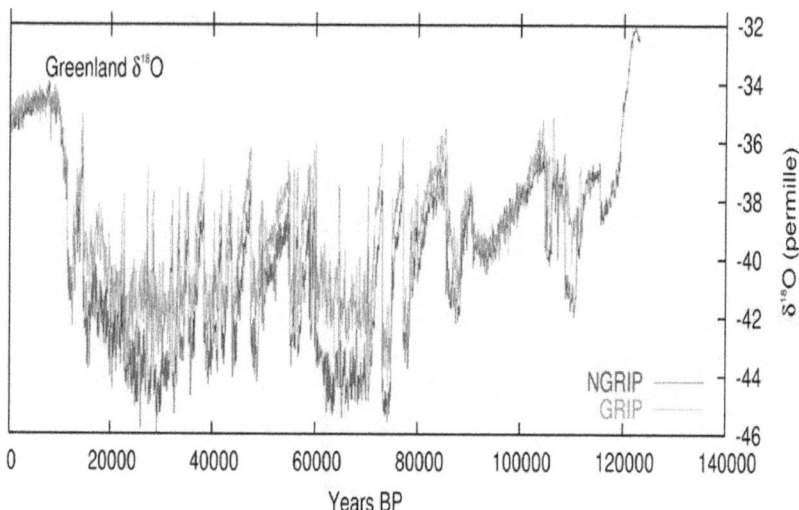

The ice core records from Greenland, that were produced from two different drilling sites, tell us both, that even during the deep glaciation period numerous events occurred that have rapidly warmed the Earth up from deep interglacial cold climates to near interglacial conditions.

These extremely large and rapid oscillations make perfect sense in the eclectic world of the Primer Fields where that come to light as miniature interglacial events that are caused by the Sun becoming periodically active again for short intervals, with the Primer Fields becoming re-established for as long as the conditions hold, which is totally possible in a resonating electric system.

At the beginning of the last Ice Age

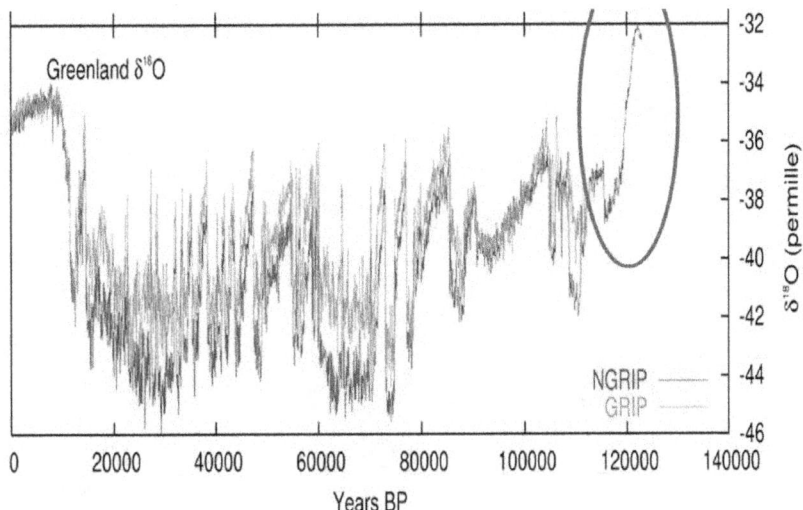

At the beginning of the last Ice Age the temperature derived from ice core samples in North Greenland dropped off steeply to about the mid-point of the deep glaciation level. This made the Earth about 20 times colder than the Little Ice Age had been.

Rapid oscillations in Greenland ice

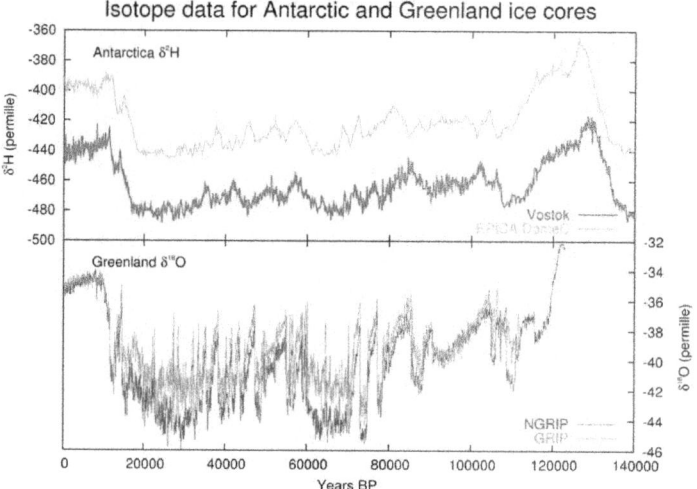

However, the sharp drop-off doesn't show up in ice core samples drilled from the ice in Antarctica, nor do the rapid oscillations show up that span the entire glaciation period, which are clearly evident in the Greenland ice core samples.

The Sun can alternate on and off states

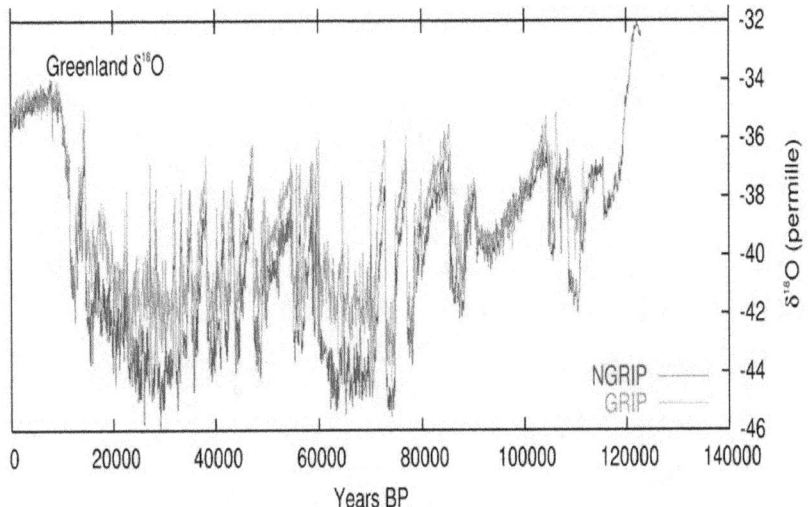

Does this mean that the rapid oscillations did not occur, or might be local occurrences? It is tempting to assume this, because without the Primer Fields theory that makes it rather plain that the Sun can rapidly alternate between on and off states, it is almost impossible to explain the large and fast climate oscillations that the Greenland ice core samples tell us of.

Greenland ice is much more sensitive

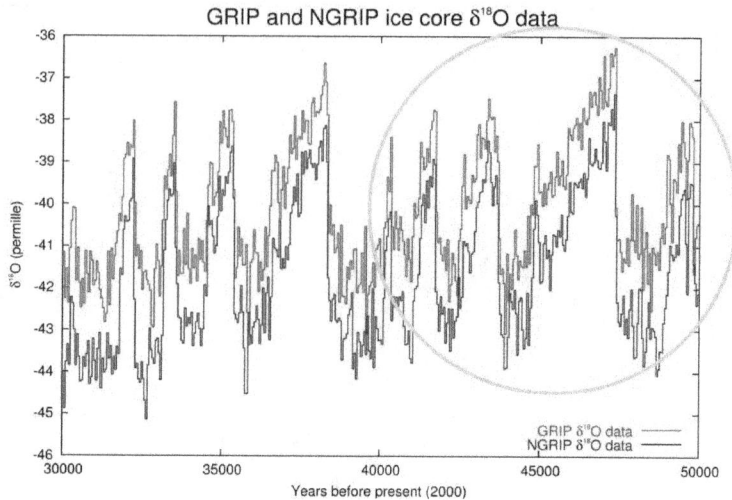

But with the Primer Fields theory considered, the enigma resolves into simply a series of events in which the Sun is actively powered for a brief span, followed by a longer span of the Sun becoming inactive again.

The Greenland ice is much more sensitive, and able to preserve these fast fluctuations, than the ice in Antarctica.

Antarctica, being an ice desert

Antarctica, being an ice desert, one of the driest spots on earth, it would be naturally less responsive to short-term variations of the type that we would expect to see when the Sun switches from its powered state to its non-powered state, where it shines only 'dimly' with its internally stored up energy.

Oxygen isotope O-18 ratio is temperature sensitive

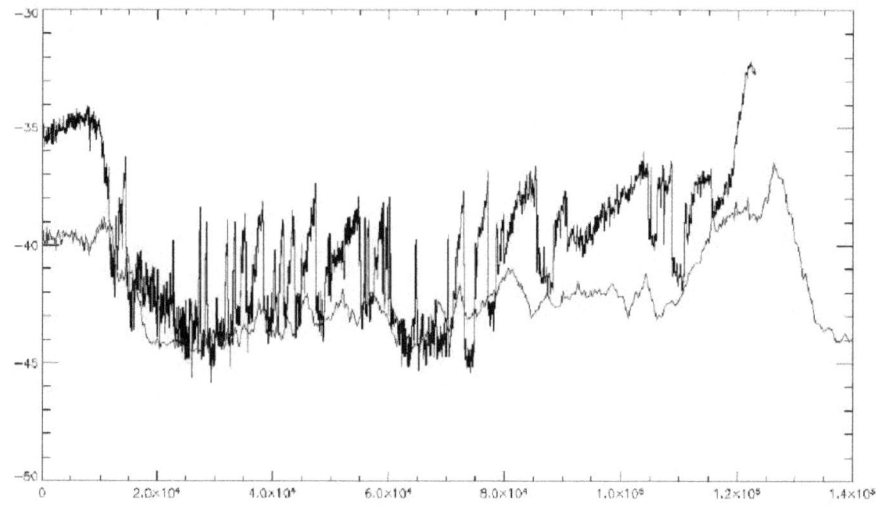

The type of rapid fluctuation that would be indicative of on-off transitions of the Sun would evidently be more strongly apparent in the ice core samples drilled from the ice sheets in Greenland, which is far from being a desert.

The temperature record that is shown here is from the North Greenland project where the temperature range gleamed from the ice samples is several times larger than the equivalent in Antarctica. In both cases the temperature gradient is gleamed from the ratio of the heavy oxygen isotope O-18 in the air, or the heavy hydrogen H-2 in Antarctica. This ratio is temperature sensitive. Colder temperatures produce a greater concentration of O-18.

Antarctica the washed out major trends

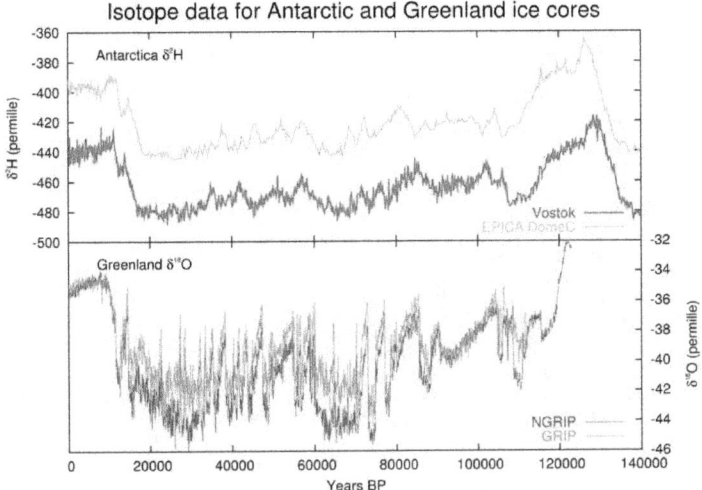

The baseline concentrations also vary somewhat between the drilling sites, both in Antarctica and in Greenland, but show the same pattern. What we see in the Greenland patterns presents strong evidence of the type that one would expect for solar on-off conditions, of which the Antarctic ice shows only the washed out major trends.

Rapid fluctuations in the Greenland ice

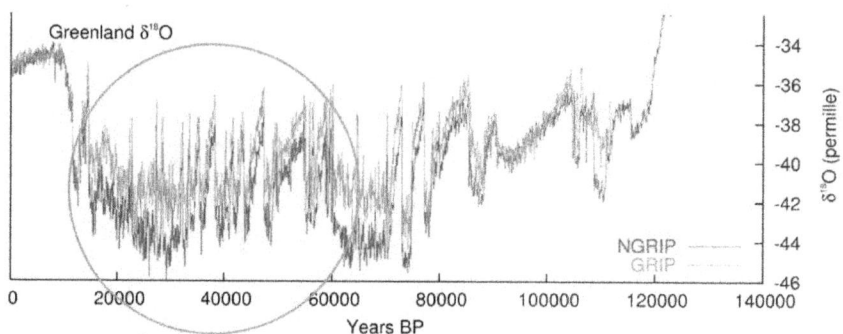

Dansgaard–Oeschger events evident in the Greenland Ice Core samples

The rapid fluctuations that are detected in the Greenland ice, are called the Dansgaard Oeschger oscillations. These are transitions from deep glaciation conditions, almost all the way back to interglacial conditions.

Dansgaard Oeschger oscillations

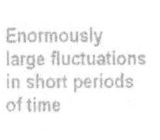

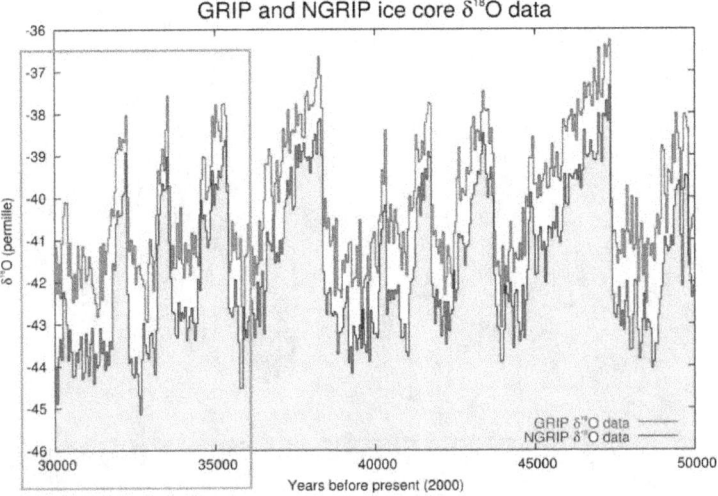

These are enormous fluctuations. They tell us that the Earth heated up from full ice age conditions to near interglacial temperatures in twenty to thirty years, and then cooled down again, gradually, to deep glaciation conditions, in a few hundred years. This rapid warming, and gradual cooling, is evident in all of the big Dansgaard Oeschger oscillations. All told, 25 of these big fluctuations have been recognized.

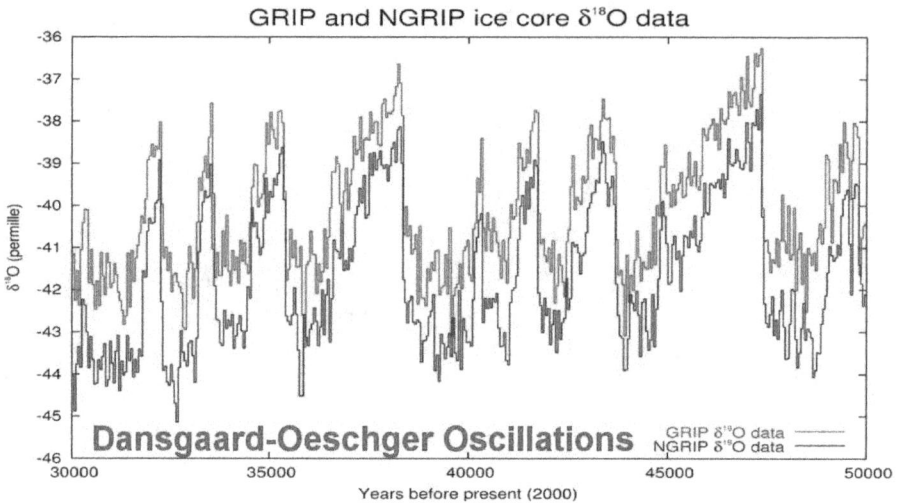

In some cases, where the interval is long - where the gradual cooling spans a longer period - we see evidence of small, sharp, upwards spikes along the way, suggesting that numerous short bursts of the powered state of the Sun have occurred that have caused a periodic re-warming of the Earth, and also of the Sun itself, internally.

The evidence suggests that the last entire ice age, and those before, were created by a long series of the power-off state of the Sun interspersed with short periods of the Sun being fully powered. It appears that whenever the Primer Fields are established, the Sun is actively powered to roughly its full potential, and that both the Sun and the Earth cool down during the longer powered-off periods.

Giant red sprites

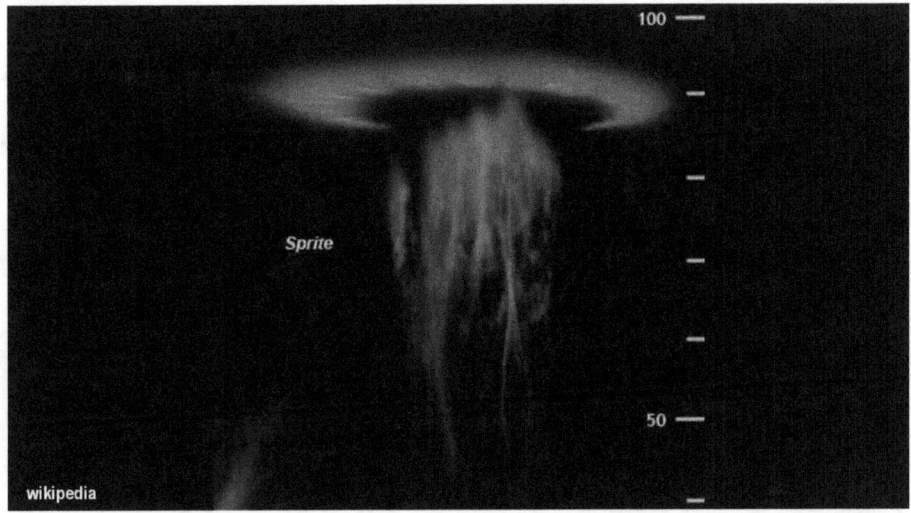

It may seem irrational to assume that such a giant system, as that which feeds our sun, can turn on or off in the space of the day. However, in electric systems, such rapid transitions are totally natural and are readily observed in the natural environment. When storm clouds reach high into the atmosphere, an electric field becomes established at times that extends into the stratosphere and causes a plasma-flow connection. When this happens the basic pattern of the Primer Fields appear in the sky. They flicker on strongly in the shape of giant red sprites, and then vanish just as fast. They rarely last for more than a second. The Primer Fields that activate our Sun, evidently can 'flicker' on and off in a similar rapid manner.

After the Sun turns off

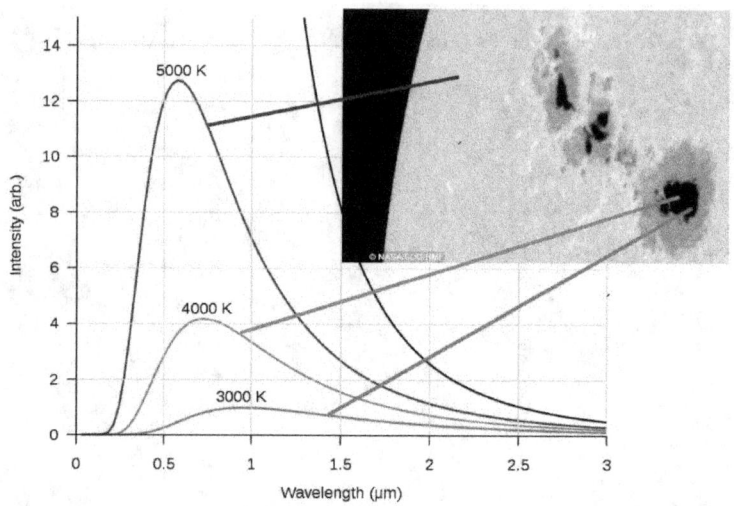

What we see in the resulting relationship between the on-and-off periods for our Sun has an enormous impact on what will happen to the agriculture that we depend on, the moment that the turn-off happens that marks the start of the next Ice Age. While we may see a gradual cooling of the climate of the Earth after the Sun turns off, the effect of the cooler Sun will have an immediate impact on agriculture. Not only will there be radically less energy available for the chlorophyll of the green plants to function, but the radiation spectrum will also shift away from the wavelengths of the visible light that are critical for the absorption by chlorophyll, without which plants cannot grow.

When the Sun turns off, we will see an immediate 70% reduction of the total energy coming from the Sun, and an immediate shift of the energy profile towards the red. We can compensate for the energy loss by placing our agriculture into the tropics, enhanced with artificial lighting, and by placing some parts of it directly into indoor facilities with 100% artificial environments.

Absorption spectrum of chlorophyll

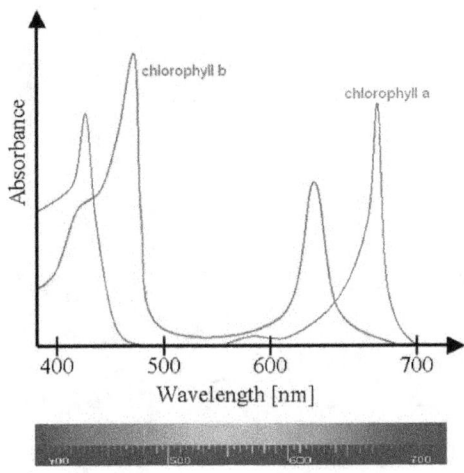

While the Sun becoming inactive poses some challenges, the challenges are not insurmountable. As you can see, both types of chlorophyll get most of their energy from the shorter wavelengths below 500 manometers, which the Sun provides even less of when it dims down. However, since the absorption spectrum of chlorophyll is narrow and specific, only small amounts of energy are required when the lighting is tuned to the absorption bands. At the present time, only 2% of the solar energy received is actually utilized by the plants. The entire amount then, can be provided with relative ease with the use of nuclear or more advanced types of electric power systems.

When the Sun enters its off-state

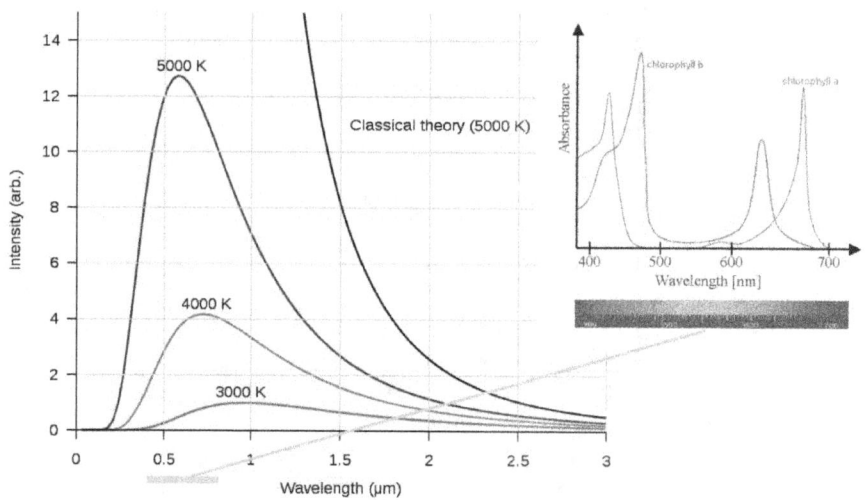

This means that when the Sun enters its off-state, that is its inactive residual heat-state, the remaining sunlight will be essentially useless for most types of agriculture existing today. This means that all the new agricultural platforms that need to be built to maintain our food supply, will have to be in place and be operating, before the day the Sun becomes inactive. This is the new reality. The transformation of the Earth will happen almost 'instantly,' possibly in the span of just a single day, or less.

When the solar off-transition happens, and it will happen, all of the temperate-zone agriculture, where presently nearly all of the world's food is produced, will be 'instantly' disabled. The transition won't happen gradually. It will happen 'instantly.' It will happen without warning. And when it happens, the consequences will begin immediately, on the very day. That's what we have to be ready for. That's what we must prepare for. The larger climate transition that unfolds along the way, in which the snows no longer melt, but increase, as big as this will all be, will actually be of lesser importance then.

Agriculture afloat on the equatorial seas

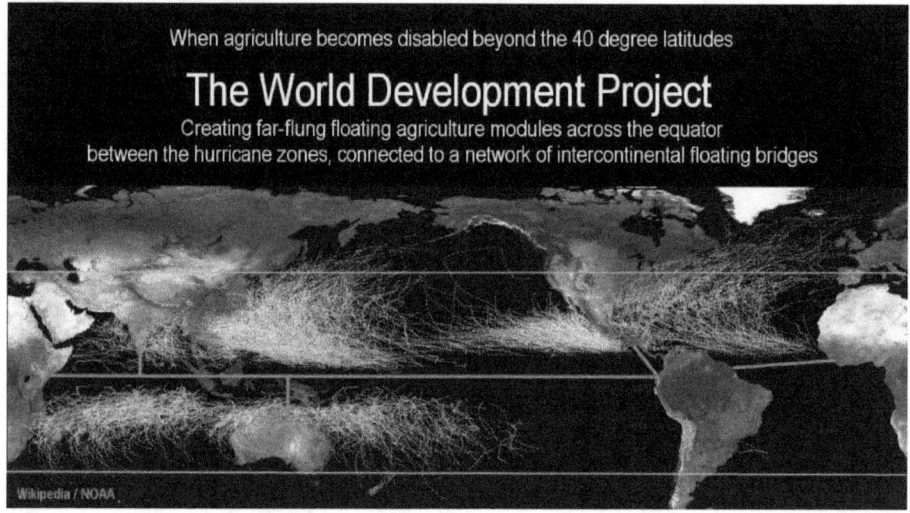

When agriculture becomes disabled beyond the 40 degree latitudes

The World Development Project
Creating far-flung floating agriculture modules across the equator
between the hurricane zones, connected to a network of intercontinental floating bridges

Wikipedia / NOAA

The impending ice age transition thus forces us to begin the greatest world development ever imagined, and to do this on a gigantic scale, like placing 90% of the world's agriculture afloat on the equatorial seas, all connected by floating bridges, and it being serviced by a new society living in floating cities along the way.
The requirements that we must meet for the future, demand us to start a new age of automated industrial production in the present - this means now and without fail - utilizing nuclear-powered high-temperature processes with the use of basalt as the feedstock. Basalt is none-corrosive and is lighter and stronger than steel, and is infinitely available. This also means that houses must be produced in automated industries, where they are produced at such a low cost in efforts that they can be given away for free as a part of the new infrastructures for living that must be produced, for the human journey to continue during the long nights of deep glaciation.

Worse than the effect of a nuclear war

If no preparations are made that compensate for the loss of the
traditional food supply that results from the instant transformation
of our planet when the Sun turns off, then most of humanity will
simply starve to death. This isn't something to aim for, is it?
For this reason, the floating agriculture will be built, with enhanced
lighting, complete with floating cities to service them. These things
will happen, because if we fail, the resulting effect will be worse
than the effect of a nuclear war. The present global food reserves
won't last for no longer than just a few months. It won't be a
pleasant thing to watch seven billion people to starve to death.
Only a few million made it through the last glaciation cycle alive.
We want to do far better this time around. And we will do better.
The Primer Fields theory opens the door to understanding what we
are up against, and what we must prepare for. This gives us an
advantage that did not exist in the earlier times, but which exists
now for the first time in the entire history of life on our planet.
Likewise, the technological power exists for this to happen, which
makes it possible for the first time too, to build the infrastructures

that we require to 'weather' an ice age with.

Where the sunlight is the strongest

To get there, our entire civilization will likely have to be rebuilt and set afloat in the tropics where land is scarce. This will happen. Shifting our agriculture into the tropics, where the sunlight is the strongest, will offset the early portion of the loss of the solar energy as the Sun is powered off. Some form of minimal artificial lighting will likely have to be added, and artificial climate control will likely have to be added even in the tropics, together with increased CO_2 concentrations for increased plant growth.

These infrastructures will require some extensive scientific advances in plant biology and in physical engineering technology for the floating new environments, and the whole thing will have to be in place, and the infrastructures be operational, without fail, before the transition happens. This means we should get started now. We still have a chance to get this critical work done. We better not waste this chance before us, by doing nothing. It may be our last chance.

The next deep glaciation to begin

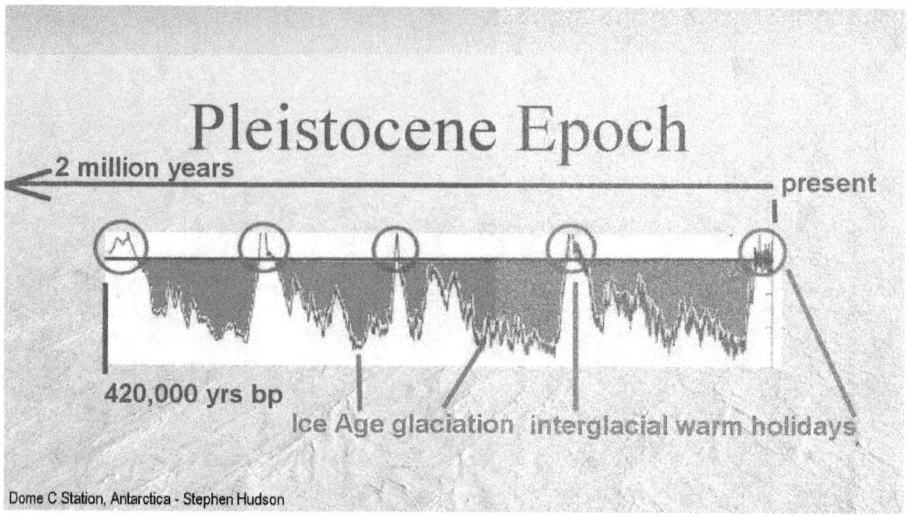

The ice age transition will happen without fail. We can count on that. We are close to the transition to the next deep glaciation to begin, though we don't know how close, close is.

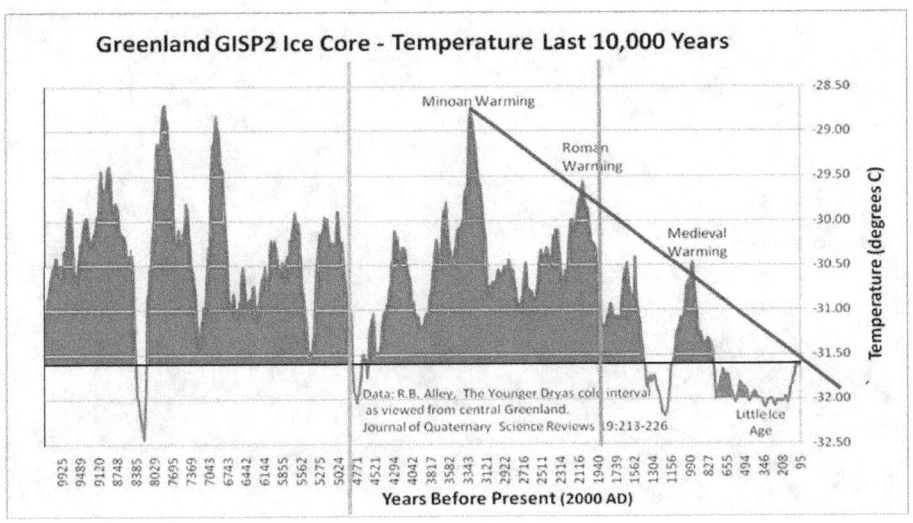

Greenland GISP2 Ice Core - Temperature Last 10,000 Years

We also know that the trend towards the next glaciation has been progressing for 3000 years already, and is accelerating. We don't know exactly where the cut-off level is, beyond which the Primer Fields collapse and the Sun turns off. It may not be far below the level of the Little Ice Age. At the present trend, we may get to this point in twenty or thirty years, or fifty at the most.

NASA's Ulysses spacecraft

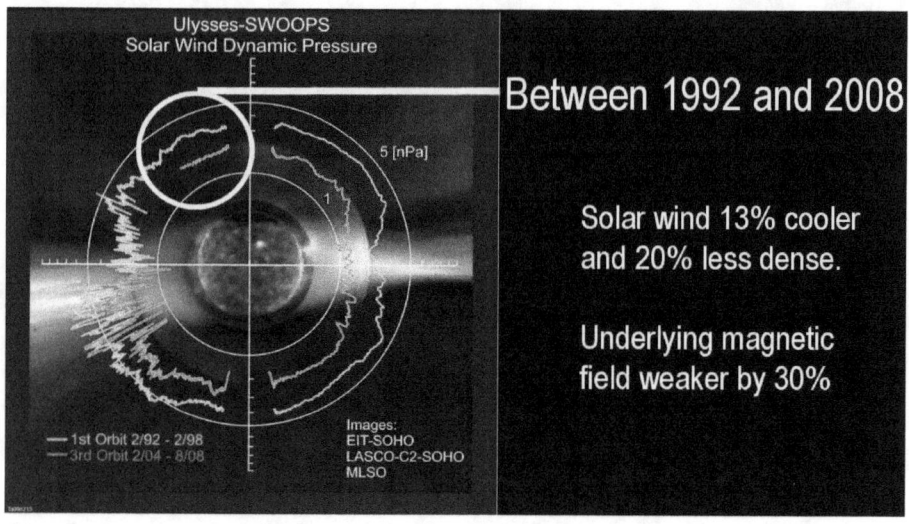

NASA's Ulysses spacecraft saw a 30% reduction in the solar wind pressure happening in just a single decade. That's not a small drop-off. And the trend is continuing. If our agriculture has not been transferred to the tropics before the high-power solar system stops, its game-over for humanity. Humanity will likely become extinct then, by the lack of food, except for a minuscule remnant of a few million that might escape the universal death. If death by starvation is what you wish for yourself and your children, then sit back and do nothing, because that way your desired fate will be assured with great certainty.

The brilliant life-giving 'fire' in the sky

Corel corp.

The stepping away from this fate begins with the recognition that our Sun would not be the brilliant life-giving 'fire' in the sky without a dense plasma sphere surrounding it from which it draws its electric power that lights up its photosphere with electric plasma interaction.

If this realization is made, half the battle is won, because everything follows, and the needed steps become logical. And this realization shouldn't be hard to make.

It has been recognized a long time ago that our Sun is not the steady state nuclear fusion furnace, which it is widely said to be, but is powered by one of the vast networks of plasma streams that pervade the galaxies and the cosmos as a whole. Since plasma has mass, the plasma surrounding the Sun becomes attracted by the Sun's gravity and interacts with its outer atmosphere, the photosphere, which thereby becomes excited to 5,780 degrees Kelvin. It is that simple.

The Primer Fields will vanish in the near future

The Primer Fields prime the environment in which the simple process unfolds for as long as the fields exist. Once you realize that this presently operating system will vanish in the near future, and that solutions are possible for humanity to move forward in spite of the dimmer Sun, you may become inspired thereby to get out of the easy chair to gain a fuller understanding of how the critical process operates on which your continued existence depends.

The Primer Fields dynamics

Part 4

The Primer Fields dynamics

Transcripts at: www.ice-age-ahead-iaa.ca

The Primer Fields dynamics

Plasma sphere around the Sun

"The Primer Fields"
lab experiments
by David LaPoint
on YouTube

The plasma physicist David LaPoint suggests that there is much more to the Primer Fields dynamics than what meets the eye. He suggests that our sun operates within a sphere of highly concentrated plasma that is generated by powerful magnetic fields, which he calls the " Primer Fields," and that the plasma sphere around the Sun is magnetically confined by these fields.
If this was not so, our sun would be but a dim speck in the sky, without the large magnetic fields acting on it that surround it with a sphere of concentrated plasma. The Earth would be a cold planet then, extensively covered with ice, as it once was 700 million years ago. But is David LaPoint correct? Do they Primer Fields really exist outside the laboratory environment? Can they be seen? And if they do exist and are visible, is it possible for these vital fields to collapse?
So, let's explore what stands behind it all.

David LaPoint uses two bowl-shaped magnets

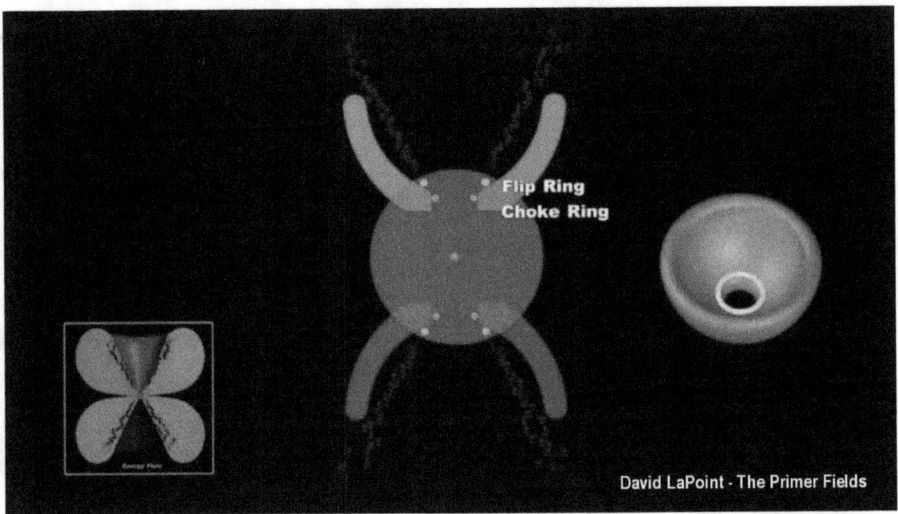

The key component of the Primer Fields theory is the existence of one or two bowl shaped electromagnetic fields. David LaPoint uses two bowl-shaped magnets of opposite magnetic polarity for his experiments conducted in a vacuum chamber.

The Zeta Pinch effect

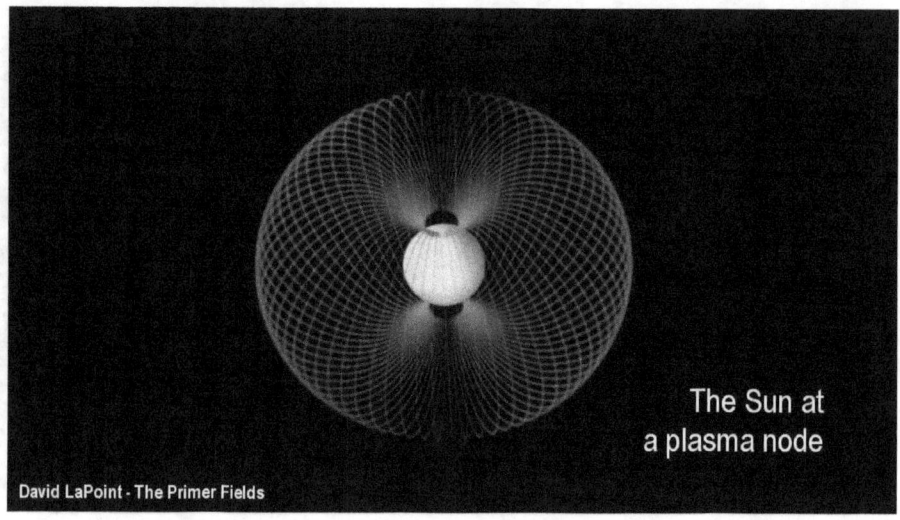

The Sun at
a plasma node

David LaPoint - The Primer Fields

In the environment of space, however, no bowl-shaped magnets hang in the sky. For the required magnetic fields to exist, they must be generated by the natural phenomenon of electricity flowing as plasma in space. And this is exactly what happens.

The term, plasma, refers to electrically charged particles that exist in free flowing form in space, primarily as protons and electrons, the stuff that atoms become made of when they are bound together. In space they are free flowing. However, flowing electricity creates magnetic fields, and by these fields the flowing electricity becomes pinched together.

When electricity is carried by two parallel wires, with the current flowing in the same direction, the wires are attracted to each other by the Lorentz force. The same happens in plasma in space where electricity is flowing freely without wires. Here the effect is called the Zeta Pinch effect.

Plasma currents in space become compressed

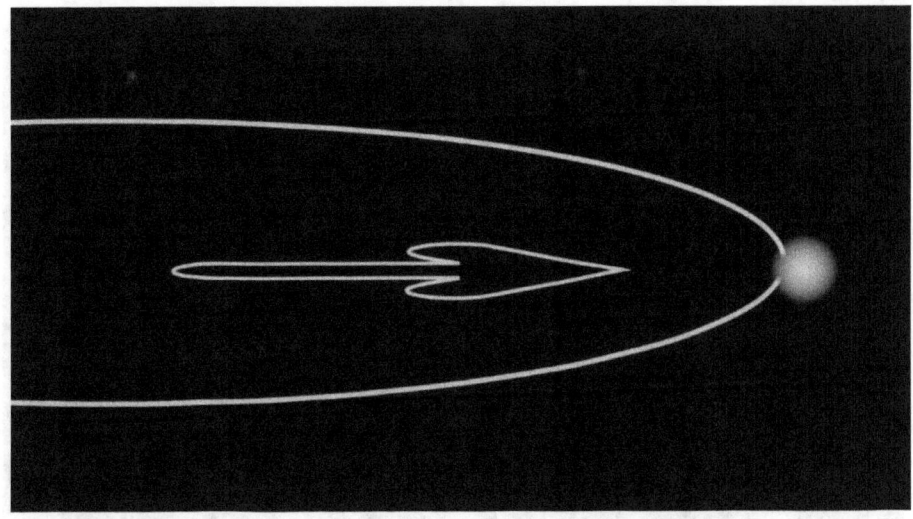

In space the flowing electric plasma particles are drawn to each other by the same magnetic forces that attract wires to each other. However, while the wires remain physically fixed, plasma currents in space become compressed into ever-smaller magnetic confinement. By the confinement the current density is increased, which in turn pinches the plasma currents still tighter and tighter, forming a bowl-shaped magnetic field in the process at the very end of the pinched plasma stream.

Electromagnetically confined 'high-density' plasma On the platform of the Primer Fields

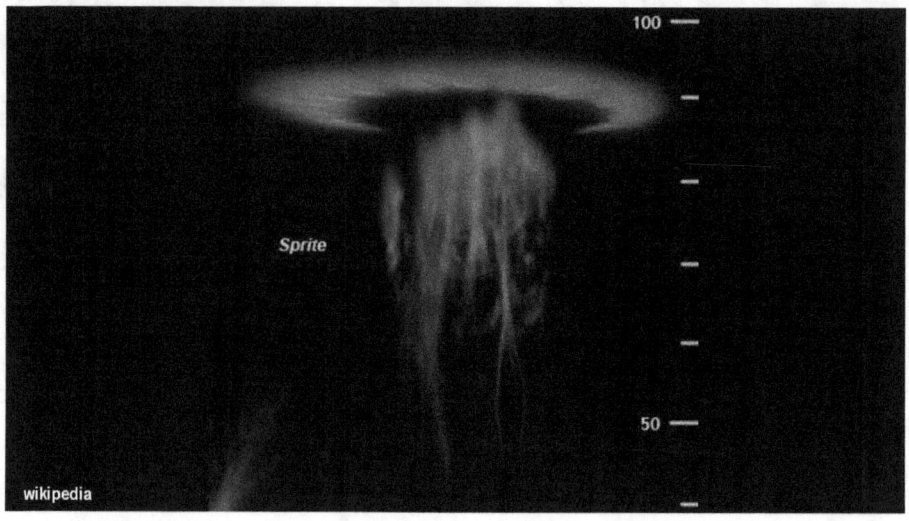

But something happens in the bowl when the plasma currents exceed a critical limit, as they converge to ever-tighter confinement. The currents and the resulting magnetic fields become unstable by the pinch effect, past a critical point. The currents become 'twisted into complex knots' whereby the bowl-shaped magnetic field that forms, opens up at the center. There, below the opening, an electromagnetically confined 'high-density' plasma stream is formed by a number of interacting effects.

On the platform of the Primer Fields the entire structure that we see that extends across 50 kilometers from top to bottom, forms in a fraction of a second. Typically, the sprite remains active for about a single second, until the plasma flow becomes too weak to maintain the Primer Fields. At the point when the Primer Fields collapse, the entire structure simply vanishes.

Interglacial period of the Sun's active time

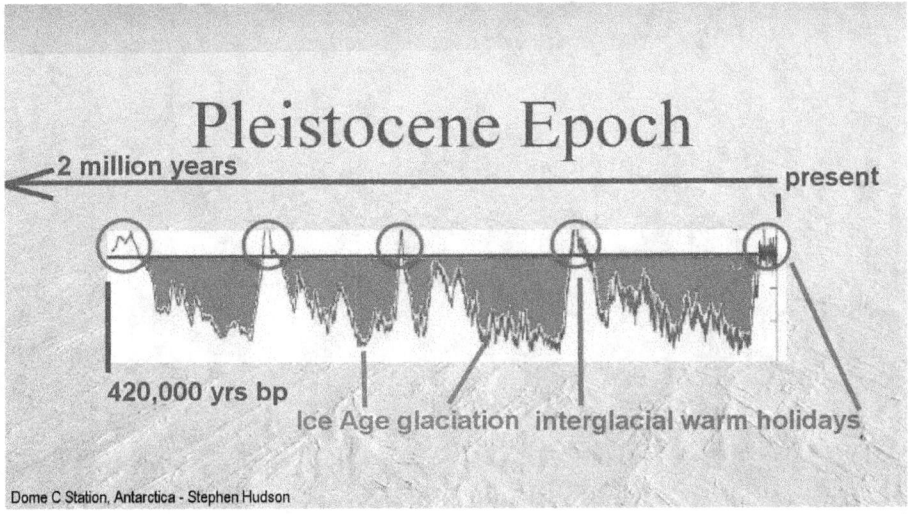

Pleistocene Epoch
2 million years — present
420,000 yrs bp
Ice Age glaciation   interglacial warm holidays
Dome C Station, Antarctica - Stephen Hudson

In comparing the sprite with the larger scale of the solar system, the sprite's one second active time is comparable to the solar system's interglacial period of the Sun's active time of roughly 12,000 years. The on-off process is the same in both cases, though different in scale.

Let me illustrate now how the process functions

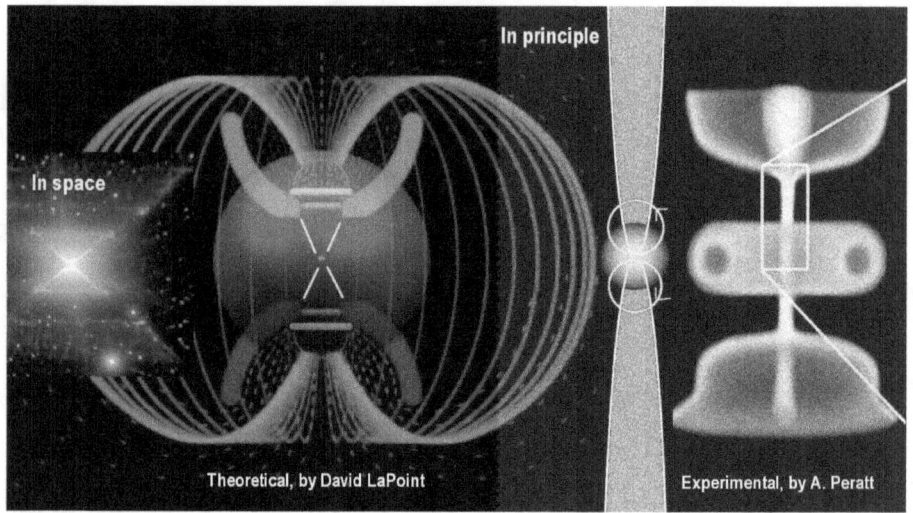

Let me illustrate now how the process functions that forms the Primer Fields.

*Dense Plasma Focus Device

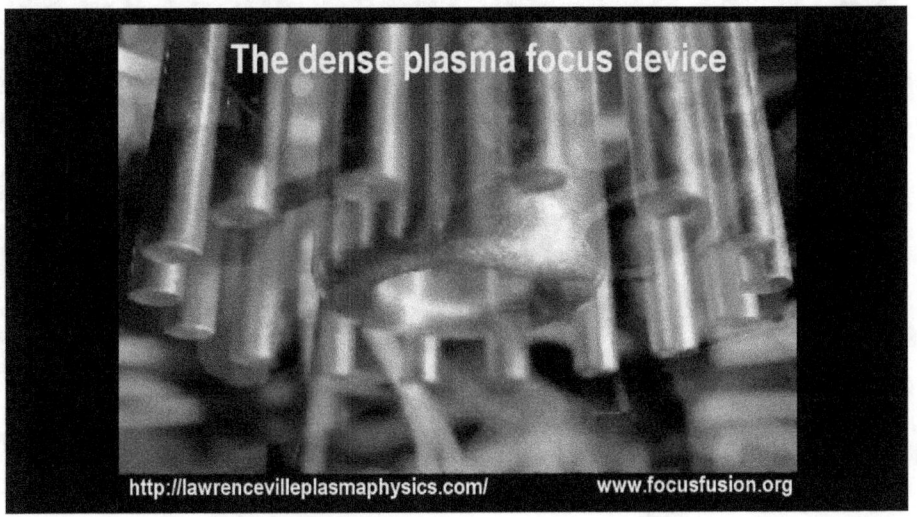

The dense plasma focus device

http://lawrencevilleplasmaphysics.com/   www.focusfusion.org

'This is best illustrated by looking at another lab experiment, that is carried out with an instrument called the, Dense Plasma Focus Device. '

A ring of electrodes

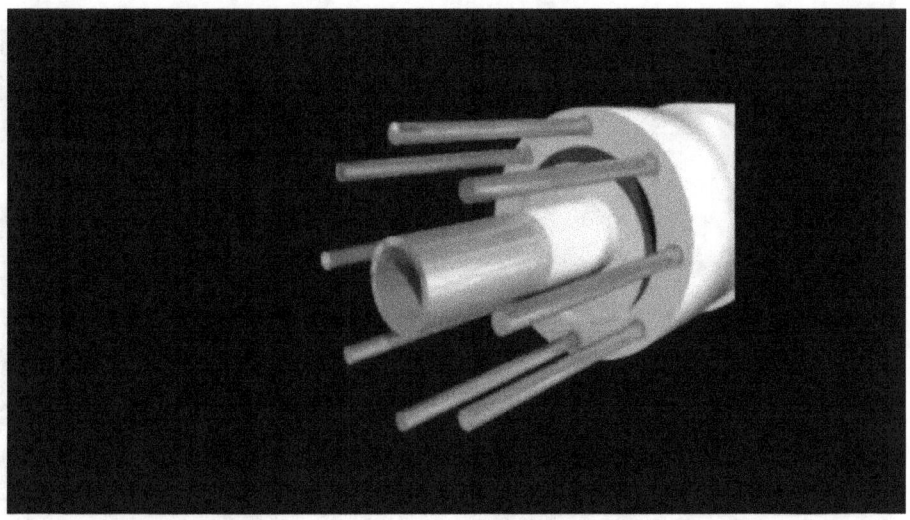

The device is made up of a ring of electrodes that surround a hollow electrode at the center.

A plasma sheet forms

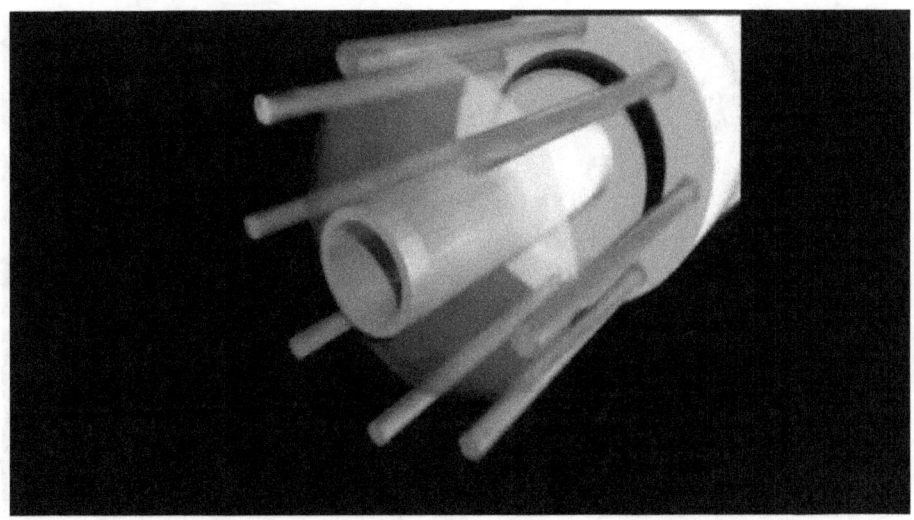

When an electric field is applied, a plasma sheet forms.

The plasma sheet

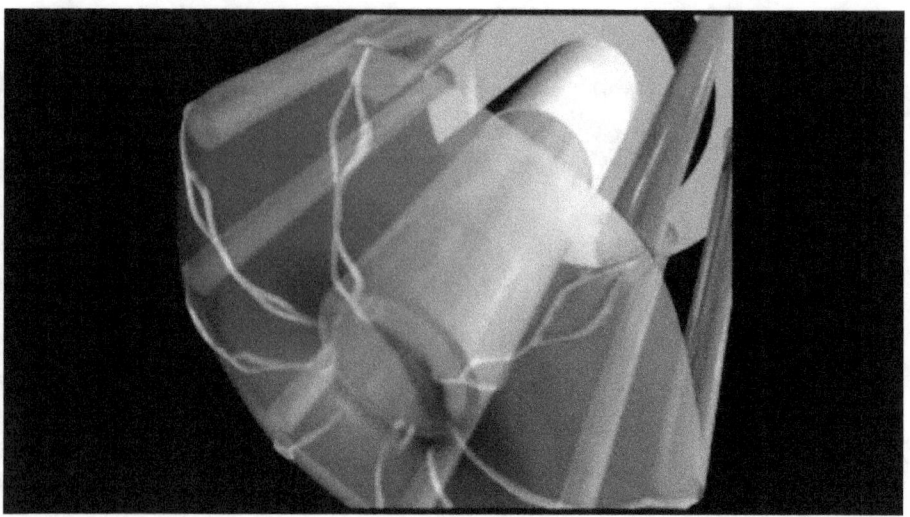

The plasma sheet is instantly drawn towards the opening of the central electrode, and is then drawn into it.

The plasma becomes extremely pinched

By it being drawn into the opening, the plasma becomes extremely pinched together.

It becomes unstable

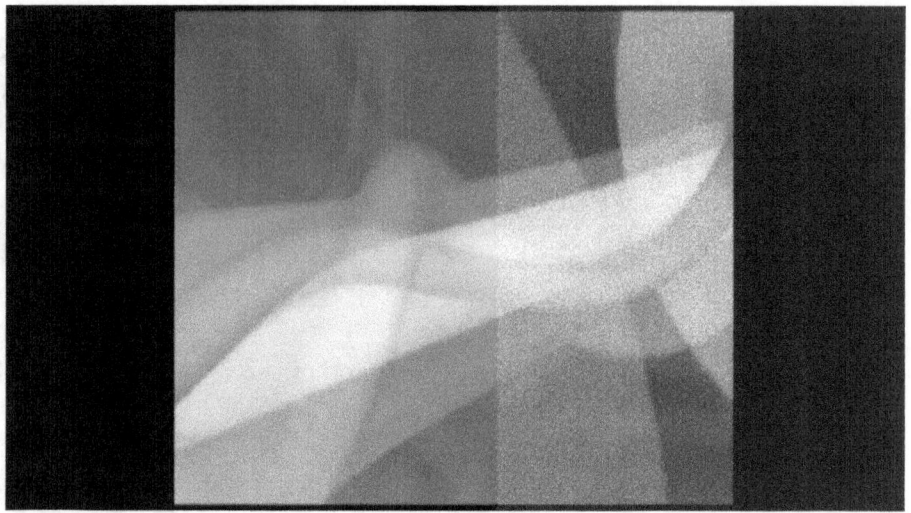

There, it becomes unstable and begins to twist.

The plasma twists itself into a spiral

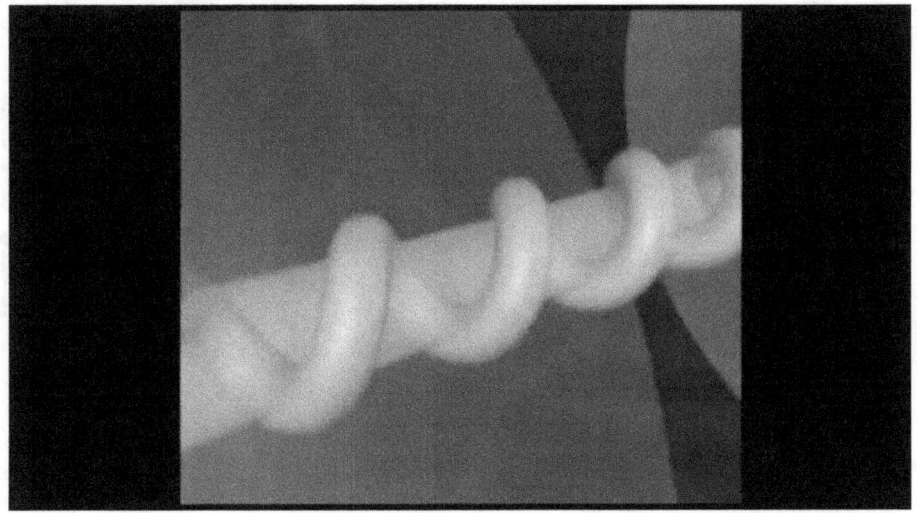

At first, the plasma twists itself into a spiral.

Then the spiral becomes compacted

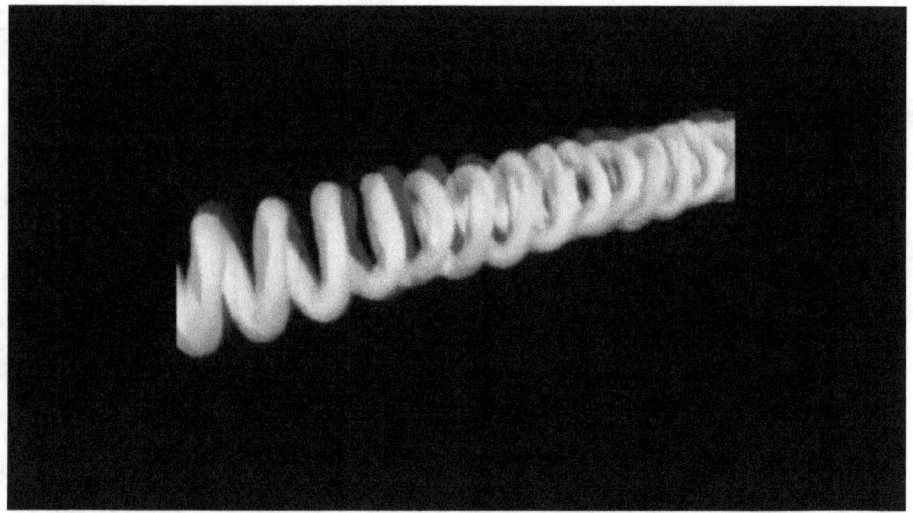

Then the spiral becomes compacted.

The more unstable the spiral becomes

The more it becomes compacted the more unstable the spiral becomes, and becomes twisted.

The twisting forms a complex knot

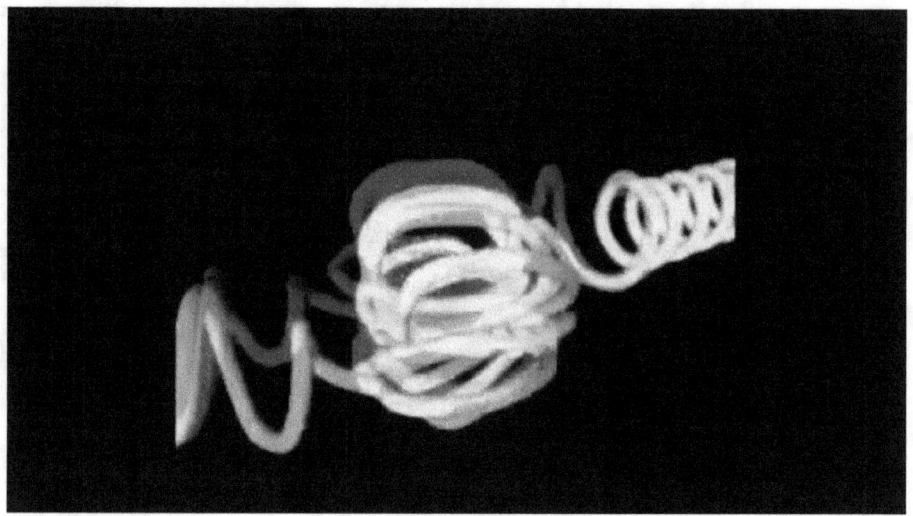

Eventually, the twisting forms a complex knot.

From a video about the Dense Plasma Focus Device

www.focusfusion.org
Animation: Torulf Greek
Editor: Rezwan Razani

Creative Commons Share Alike?

The illustrations are snapshots taken from a video about the Dense Plasma Focus Device.

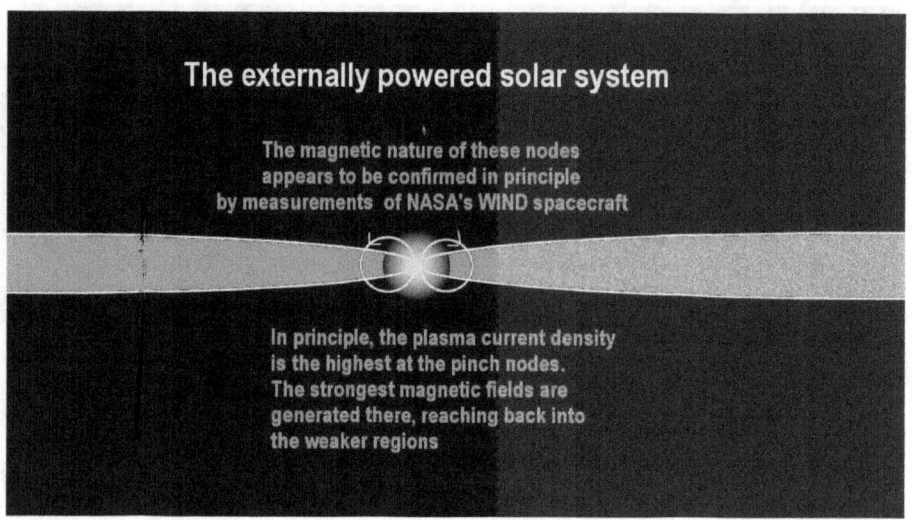

The externally powered solar system

The magnetic nature of these nodes
appears to be confirmed in principle
by measurements of NASA's WIND spacecraft

In principle, the plasma current density
is the highest at the pinch nodes.
The strongest magnetic fields are
generated there, reaching back into
the weaker regions

In space the plasma instabilities that form the complex knots open the magnetic bowls that form at the end of the concentrated plasma currents. As the process unfolds further, a high-density plasma concentration forms outside of the magnetic holes. That's where the sun is located in a solar system and is powered thereby.

When the plasma streams are too weak

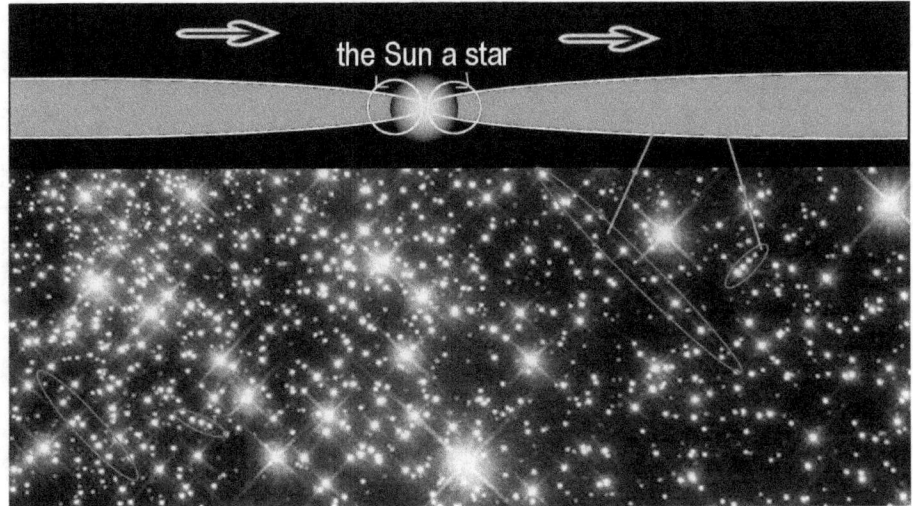

When, however, the plasma streams are too weak to cause an extreme pinch effect to happen, the plasma streams simply flow through the solar system without activating anything. The Sun thereby remains dim and not actively powered.

When the Sun is powered

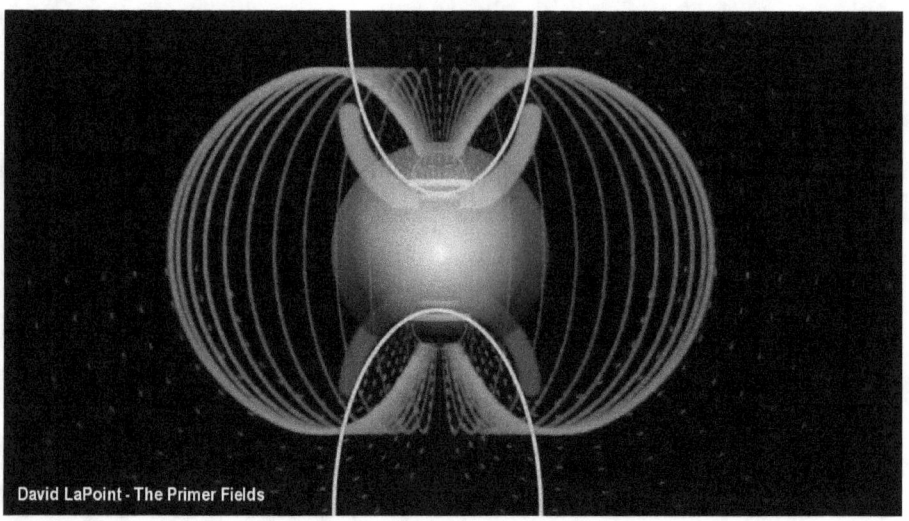

David LaPoint - The Primer Fields

When the Sun is powered by a dense plasma sphere surrounding it, the flow-through process still happens. Plasma flows out of the plasma sphere, since the Sun utilizes only a small portion of it. In the outflow another magnetic field is generated, but with the opposite polarity. The evidence suggests that all large electromagnetic structures that exist in space, when they are drawn to a sun, or on the larger scale to a galaxy, whereby the Primer Fields form, exist typically in complementary pairs.

A number of interesting effects

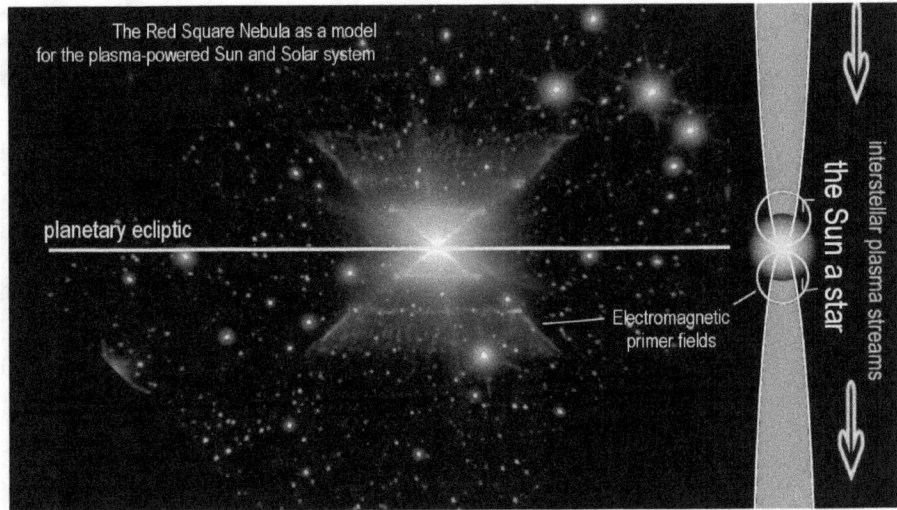

The Red Square Nebula as a model for the plasma-powered Sun and Solar system

planetary ecliptic

Electromagnetic primer fields

the Sun a star

interstellar plasma streams

Between the two giant complementary electromagnetic bowls, a number of interesting effects come to light with interesting principles that are critical for the overall dynamic interactions.

In the narrow space between

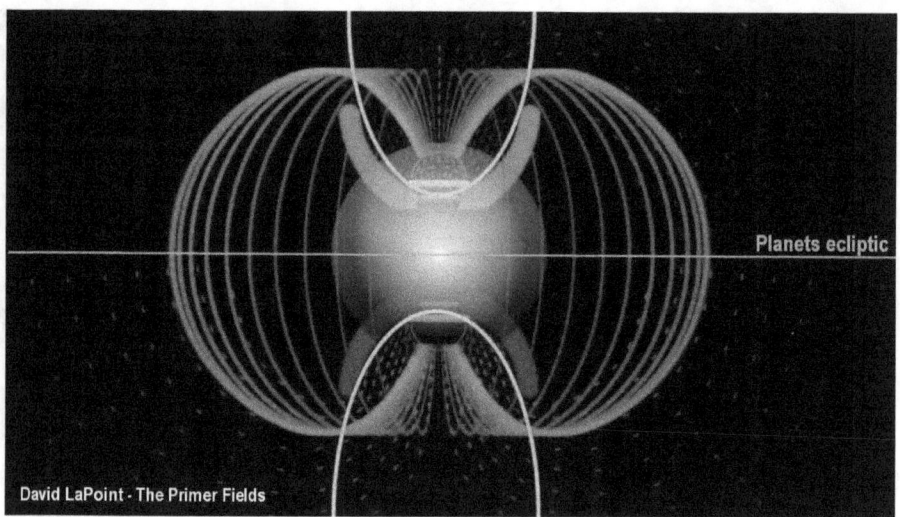

The magnetic fields are the strongest at the focal point of the bowl-shaped magnetic structures. In the narrow space between the two complementary structures lies typically a solar system with a sun, or several suns, on the central axis, and with the planets orbiting on a thin ecliptic plain in the space between the two electromagnetic bowl-type fields.

Two bowl-type electromagnetic fields work together

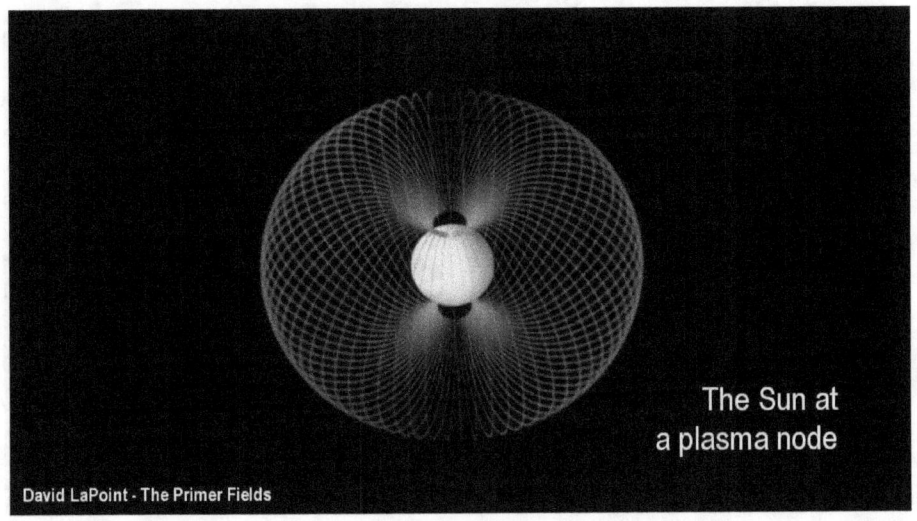

The Sun at
a plasma node

David LaPoint - The Primer Fields

While the two bowl-type electromagnetic fields are each a separate entity, they work together functionally as a whole.
Their function is to concentrate the plasma flows that pervade all space and focus them into a tightly confined sphere in which our sun is located and is powered by it.

A polarity flip point

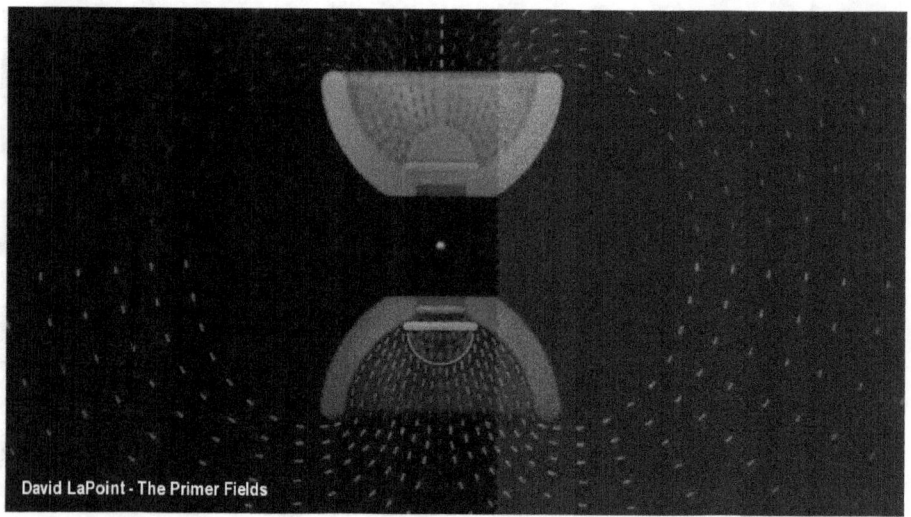

The remarkable concentration is accomplished in the laboratory by a set of two bowl-type structures, shown in red and blue, facing one-another with opposite (complementary) polarities.

The small point in the middle between the two magnetic bowl structures is where a polarity flip point is located. The flip point appears to be responsible for flipping the polarity of the Sun's magnetic field with every solar cycle.

The location of the flip point moves slightly when one of the two bowl structures becomes weaker than the other. This effect causes the polarity of the magnetic field of the Sun to assume the dominant polarity, and thus flip with the 11-year solar cycles that are simply resonance cycles between the complementary magnetic structures.

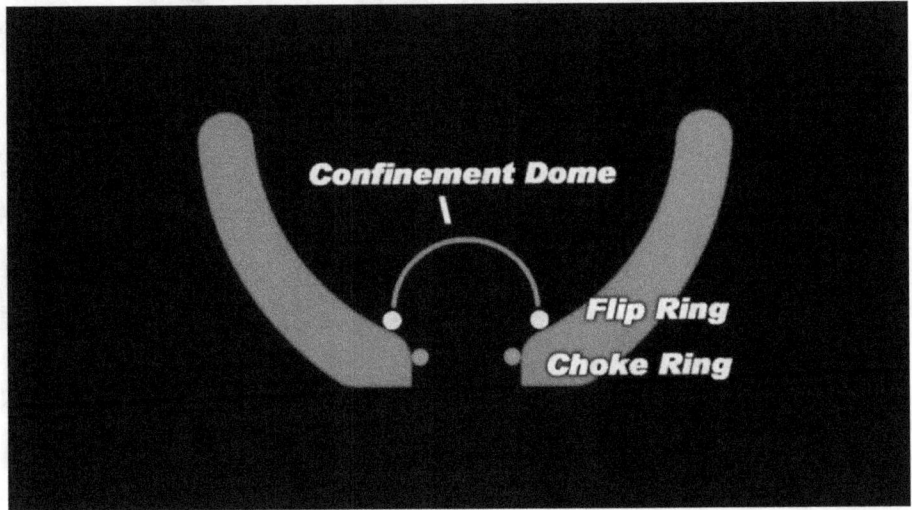

The plasma concentration that is required for the Sun to function is dynamically produced in the illustrated structure by the interaction of its three functional magnetic elements that are structured around the respective hole in the magnetic bowls.

Each has a specific function to fulfill

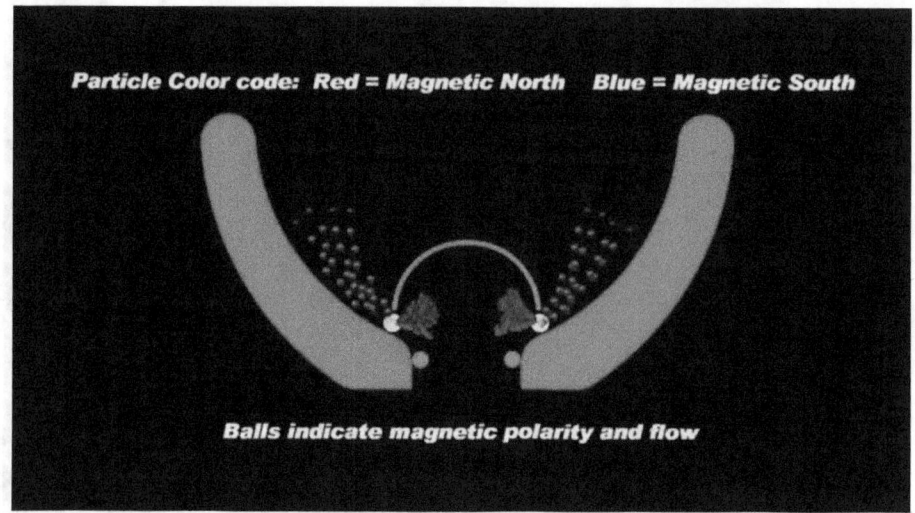

Each of the three structures has a specific function to fulfill. The flip ring flips the orientation of plasma. It flips it under the magnetic confinement dome, while the choke ring below helps to keep it there.

The flip ring

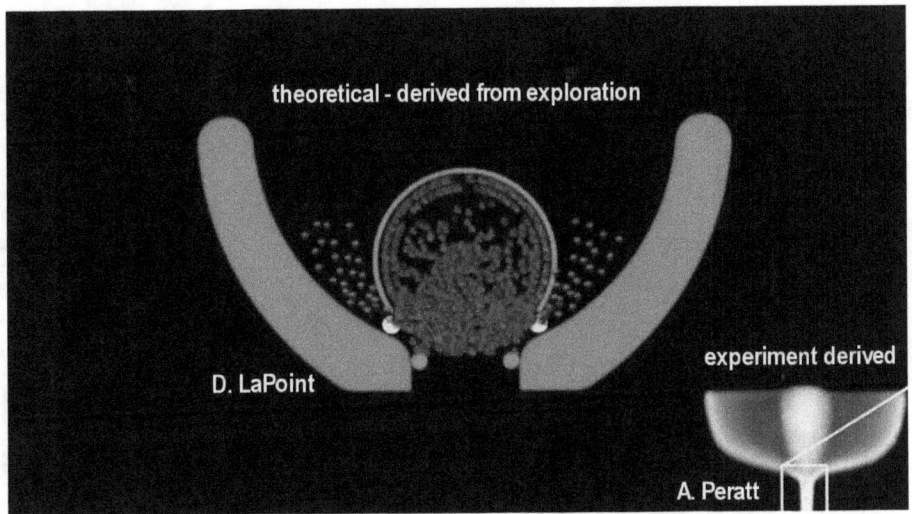

The plasma particles that flow into the big red bowl are flipped upwards as they pass the flip ring (yellow). They are collected together there into a massive accumulation. The plasma particles become concentrated by this process. The concentration creates a 'high-pressure' environment.

The magnetic choke ring

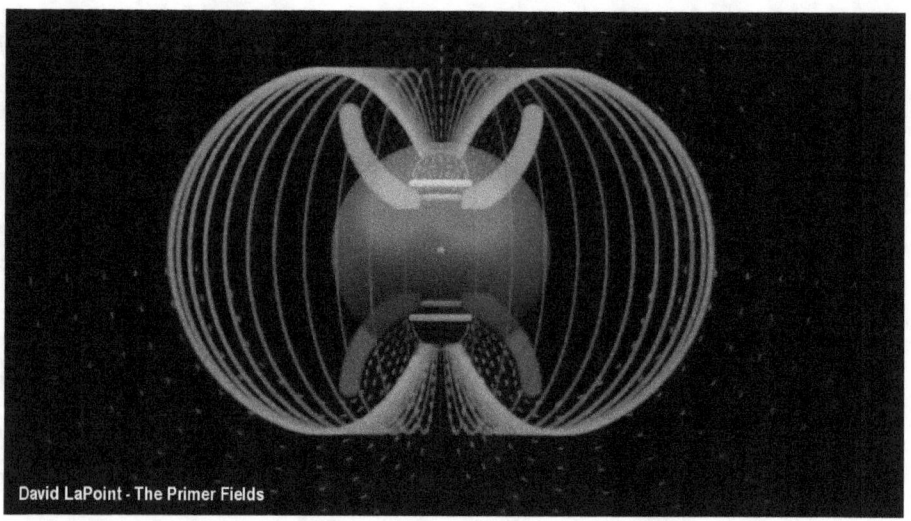

The magnetic choke ring, within the opening of the magnetic bowl, focuses the 'escaping' plasma flow into a tightly concentrated stream beneath the hole. In some lab experiments, the focused stream expands into a sphere.

The out-flowing stream

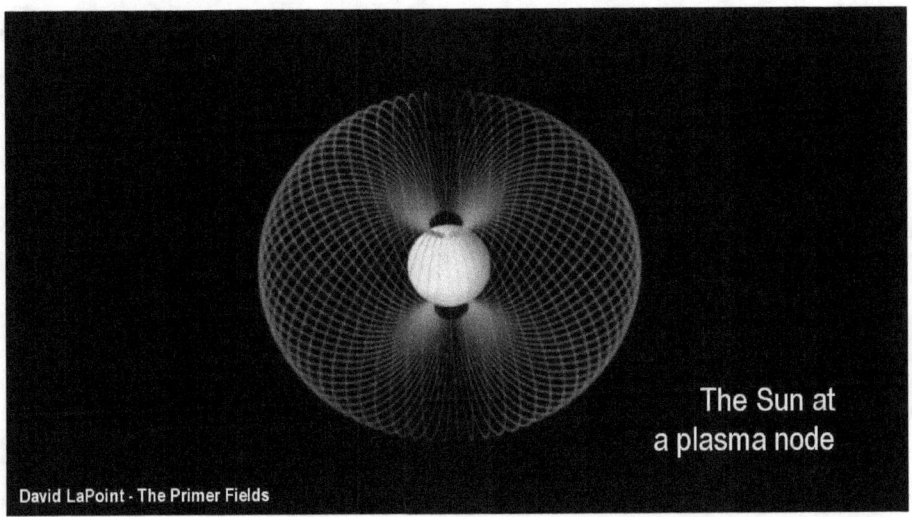

The Sun at
a plasma node

David LaPoint - The Primer Fields

When plasma is drawn out of the sphere in the flow-through process, a complementary bowl structure is formed with opposite orientation and opposite magnetic polarity. In this structure the out-flowing stream of lesser density is drawn into the bowls where expands in the reverse process, reverting back to the 'normal' density of the prevailing plasma stream.
The process that is illustrated here can be verified with laboratory experiments.

A high-power plasma experiment

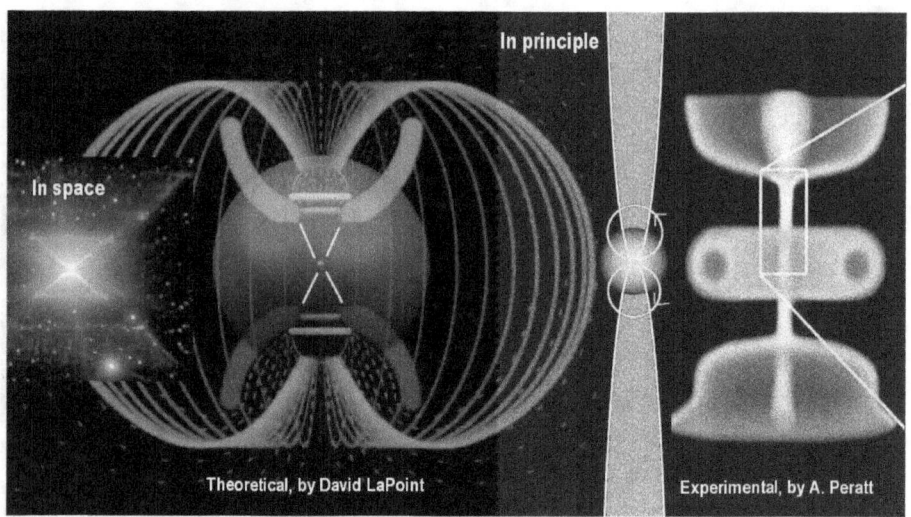

In a high-power plasma experiment of the type that is conducted at the Los Alamos National Laboratory, the complementary bowl structure that the plasma currents form by their own interaction, is clearly visible. Also the containment dome that forms inside the bowls is visible in the experiment that is illustrated here.

In this particular experiment the magnetically focused plasma stream that flows between the bowls, did not form a sphere, for which a catalyst would be required, but it did form a distinct plasma ring around the focused stream, centered between the bowls.

Archetypal drawings

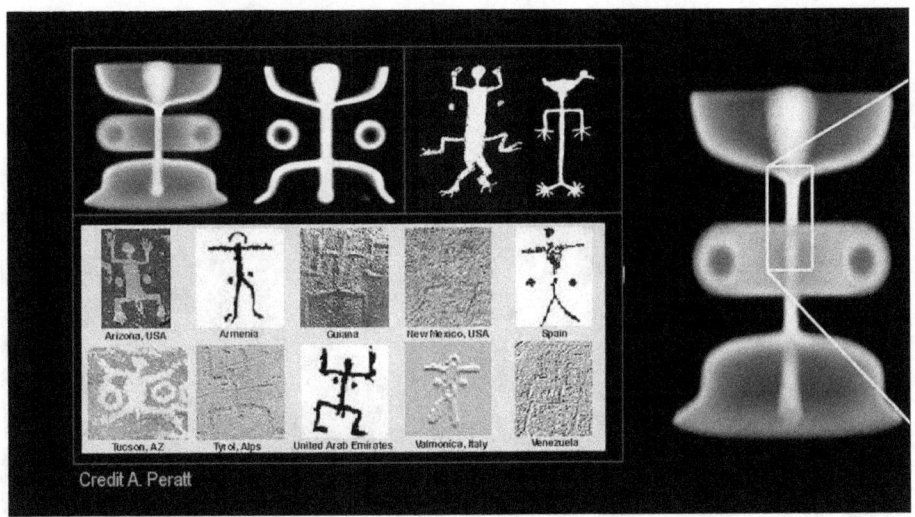

Credit A. Peratt

Evidence exists that the lab-created shape of plasma formed by the Primer Fields, was visible in ancient time in the sky. Archetypal drawings collected from widely separated regions on earth, show a remarkable similarity of their design with the lab-created plasma formations. The similarity suggests that the complementary plasma bowl shapes were common occurrences at one time, appearing and disappearing in the skies like so many UFO sightings today, or like the sprites still do under special conditions.

The plasma jets

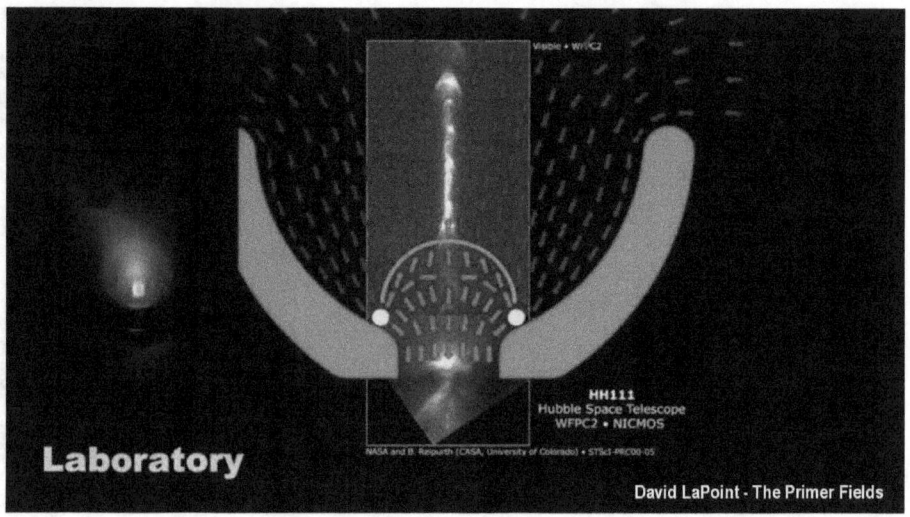

Laboratory

HH111
Hubble Space Telescope
WFPC2 • NICMOS

NASA and B. Reipurth (CASA, University of Colorado) • STScI-PRC00-05

David LaPoint - The Primer Fields

David LaPoint also discovered two more features of the Primer Fields, for which widely known evidence exists, which are the plasma jets and the magnetic flip point. He discovered that when the plasma pressure under the confinement dome becomes too great, the dome will rupture at its weakest point, by which excess plasma escapes in a burst until the rift closes up again under the resulting lower pressure. By this plasma venting process, the magnetic strength of the respective bowl structure weakens somewhat. The resulting imbalance shifts the convergence of the magnetic fields, and with it shifts a magnetic flip point that David LaPoint also discovered, forms below the opening of the magnetic bowl.

The 11-year solar cycles

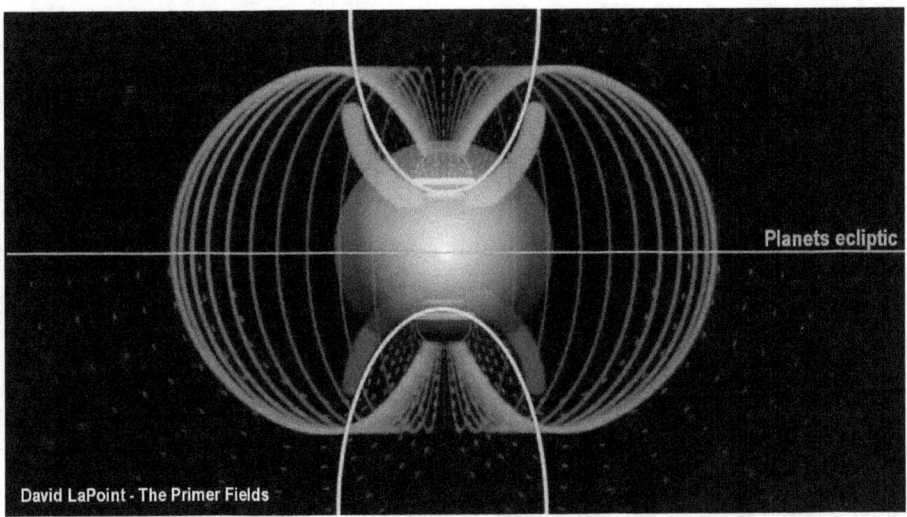

In the case of the solar system the resulting imbalance, which shifts the flip point, flips the Sun's magnetic field at the high point of the 11-year solar cycles are thereby recognized to be simply resonance cycles that oscillate between the two complementary electromagnetic structures of the solar system.

The Red Square Nebula

What the complete Primer Fields system that powers our solar system and the sun may look like in practice, can be seen illustrated in the operation of the Red Square Nebula that can serve as a model for this purpose. In this model we see all the essential features of the Primer Fields system clearly visible.

Two complementary bowl-type structures in operation

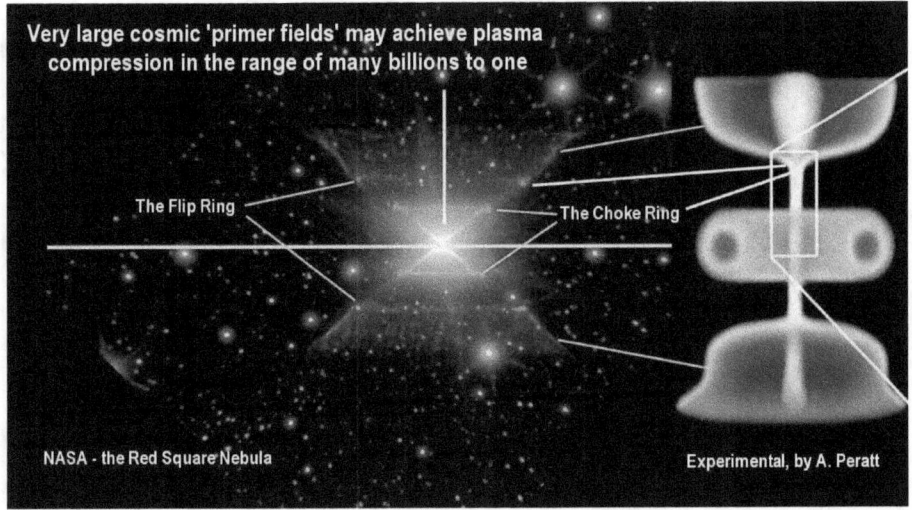

We see the two complementary bowl-type structures in operation. One concentrates the galactic plasma streams like a funnel. In the funnel we can see the flip ring, and below it the choke ring, below which a cone of concentrated plasma extends that focuses onto a sun, or a number of them. And we see the reverse happening for the outgoing plasma stream.

Observed in a laboratory plasma-flow experiment

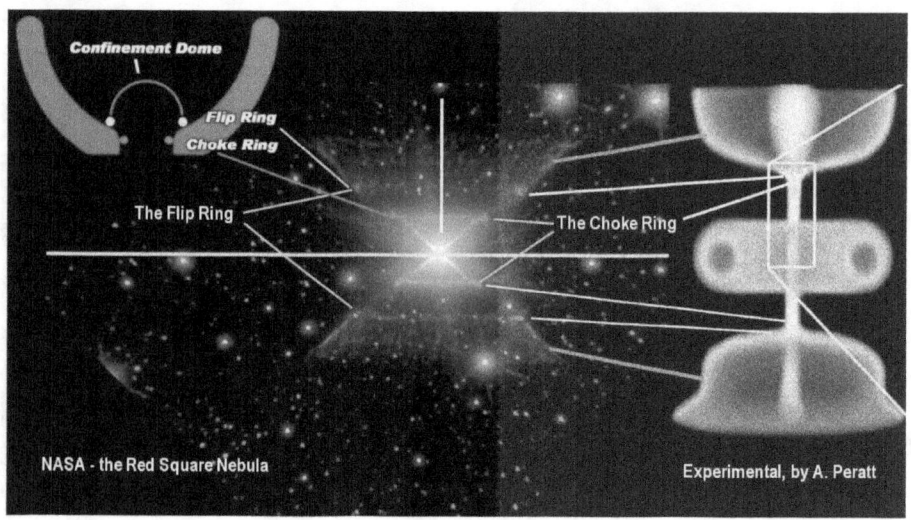

We can see all the essential features reflected here that have been observed in a laboratory plasma-flow experiment.

Our galaxy, when it is observed as a whole

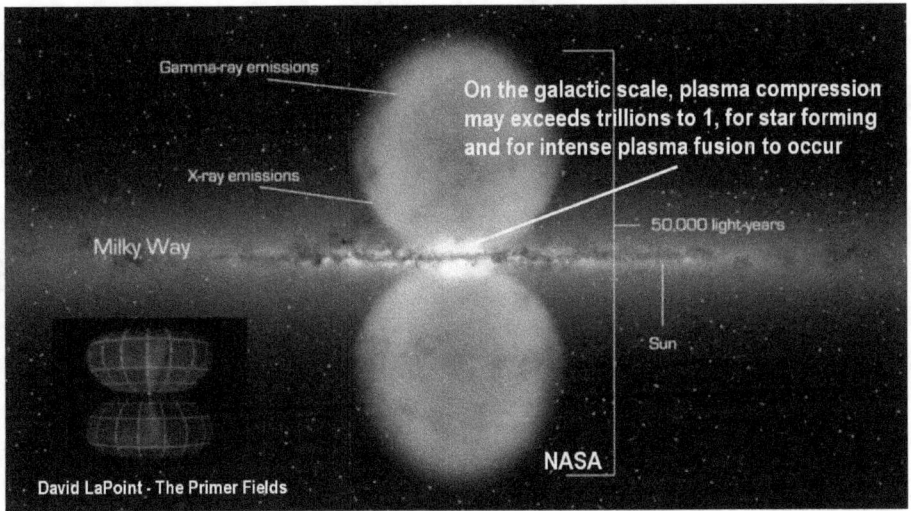

Our galaxy, when it is observed as a whole, also operates on essentially the same dynamic platform, though on a vastly larger scale than a solar system.

The center of the galaxy

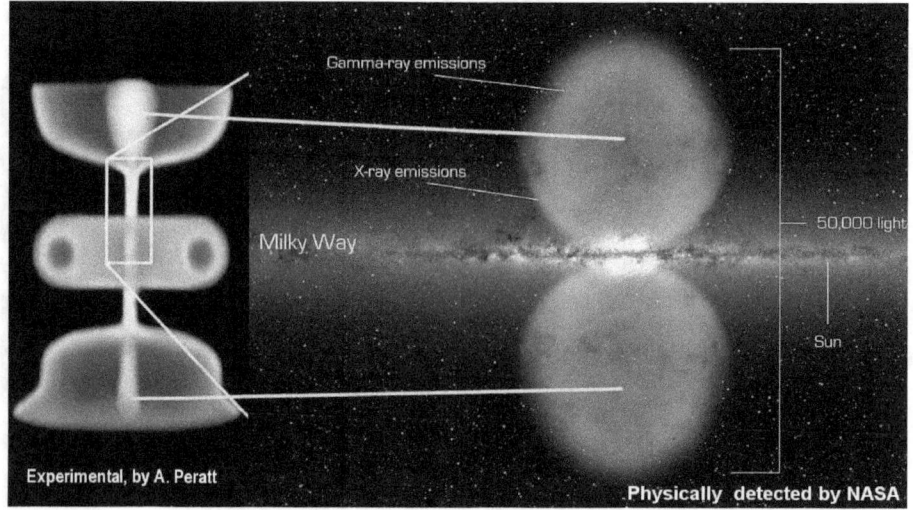

But here too, we see unmistakable evidence of two complementary electromagnetic bowls that form highly condensed plasma concentrations under their respective confinement domes. The concentration is visible in x-ray and gamma-ray emissions. We can also see the extremely concentrated plasma sphere below the confinement domes and between the two bowl structures that are indicated by the existence of the confinement domes. The concentrated plasma sphere is also the center of the galaxy.

Large intergalactic plasma streams

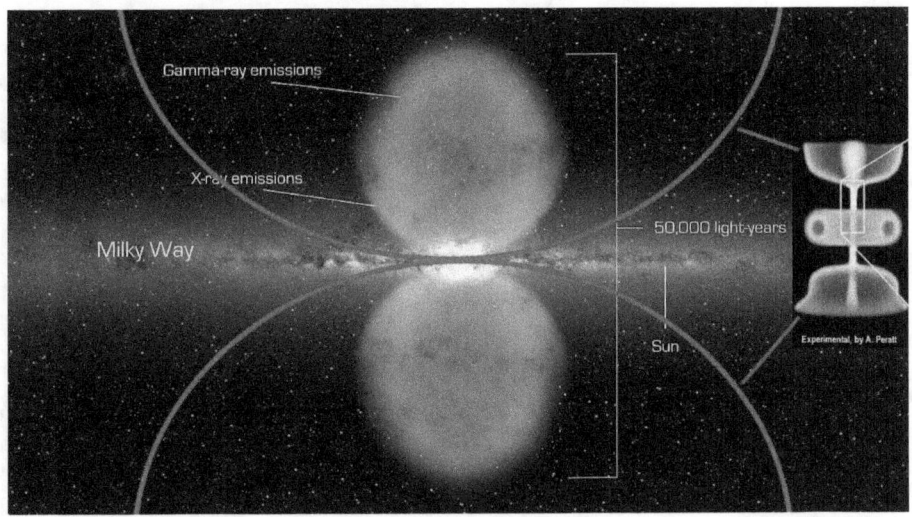

Large intergalactic plasma streams feed into and out of the Primer
Fields that form the confinement domes.

Two long intergalactic connecting streams

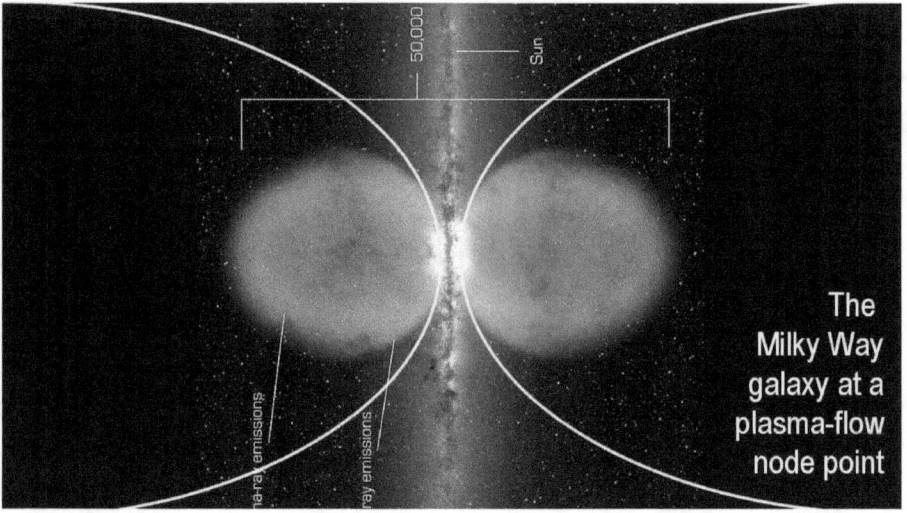

These two long intergalactic connecting streams, one incoming and one out-going, both have very long resonance cycles.

very long electric resonance cycles

Galaxies at node points

  These very long electric resonance cycles that correspond to the long distances between the galaxies, evidently affect the strength of the Primer Fields that power the galaxy, and are thereby the cause for the two long climate cycles that have been observed on Earth.

Very long electric resonance cycles

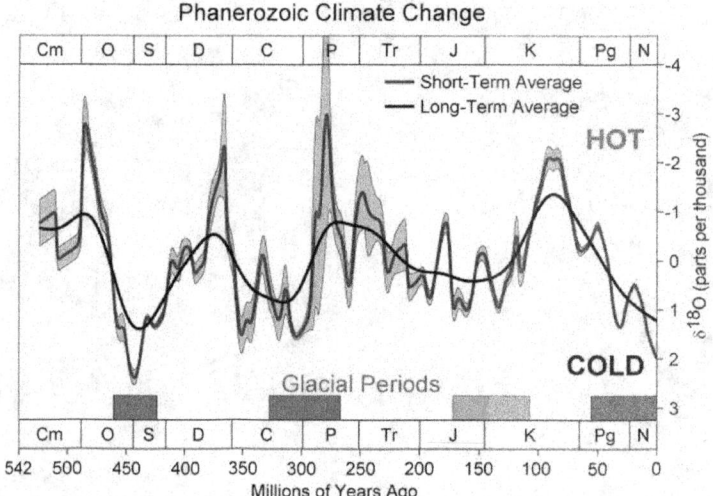

Phanerozoic Climate Change

These very long electric resonance cycles - the sixty-two-million-year cycle, and the hundred-forty-million-year cycle - show up as long climate cycles that have been preserved in sediment records that enable us to look back in time more than 500 million years.

Both near their minimum point

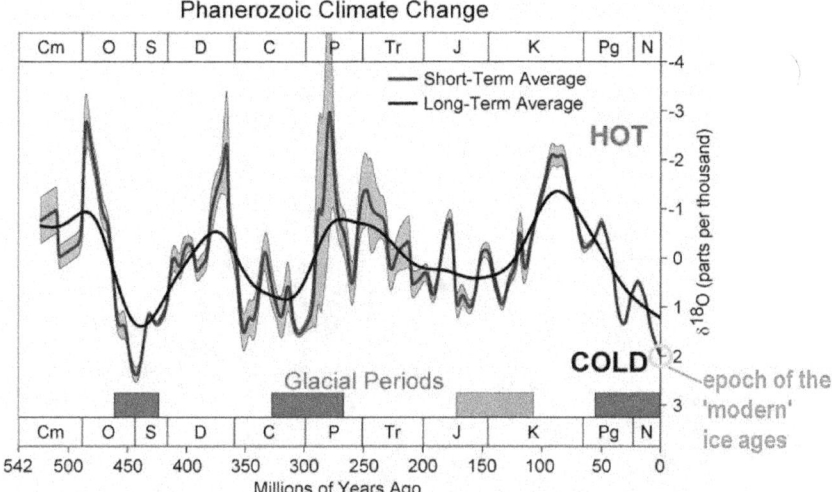

Phanerozoic Climate Change

Presently, the two very long climate cycles are both near their minimum point, whereby the weak plasma conditions have been created that have gripped the Earth for the last two million years, in which the ice ages happened.

Breakdown of the Primer Fields

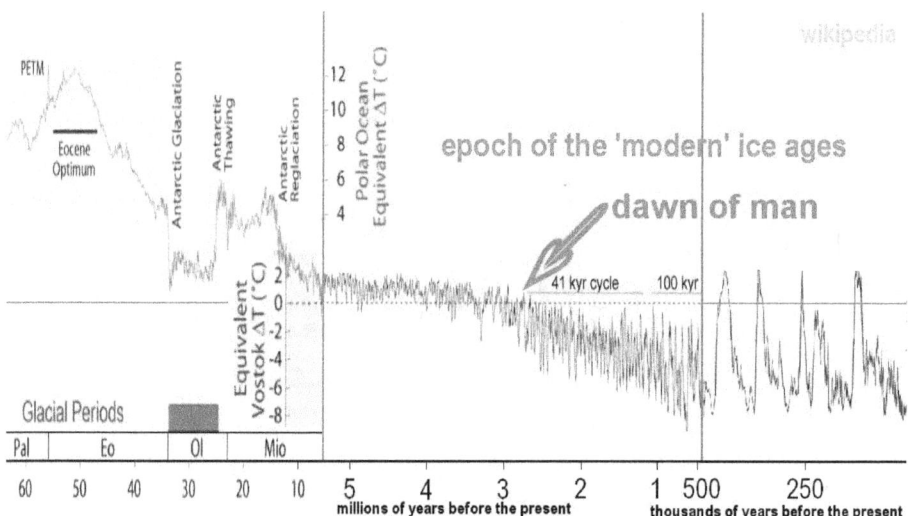

During these weak conditions that have been slowly developing for the last five million years, which are getting still weaker, the breakdown of the Primer Fields that power our solar system has become a regular occurrence during the weakest times.

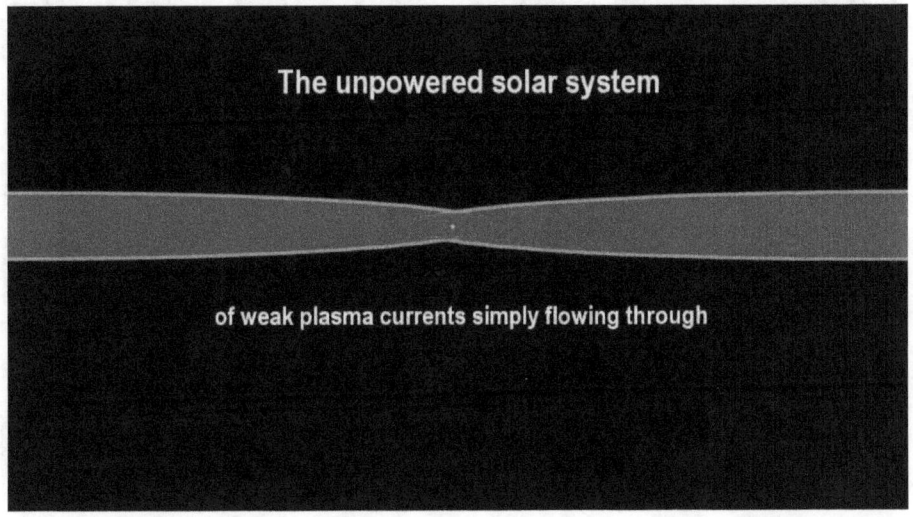

Then, when the plasma streams that flow through the solar system become so weak that they drop below a threshold, a point is reached when the pinch effect in the plasma streams is no longer strong enough to twist the plasma currents into knots to create the bowl-type magnetic structures with the void at the center that make up the Primer Fields. And so the Primer Fields cannot form. What once existed, suddenly exists no more.

The Sun becomes inactive, dim, and cold

The plasma currents become no longer concentrated then, but simply flow through the solar system without becoming focused around the Sun.

At this point the Sun simply turns off. It becomes inactive, dim, and cold. An Ice Age begins.

The Pleistocene Epoch

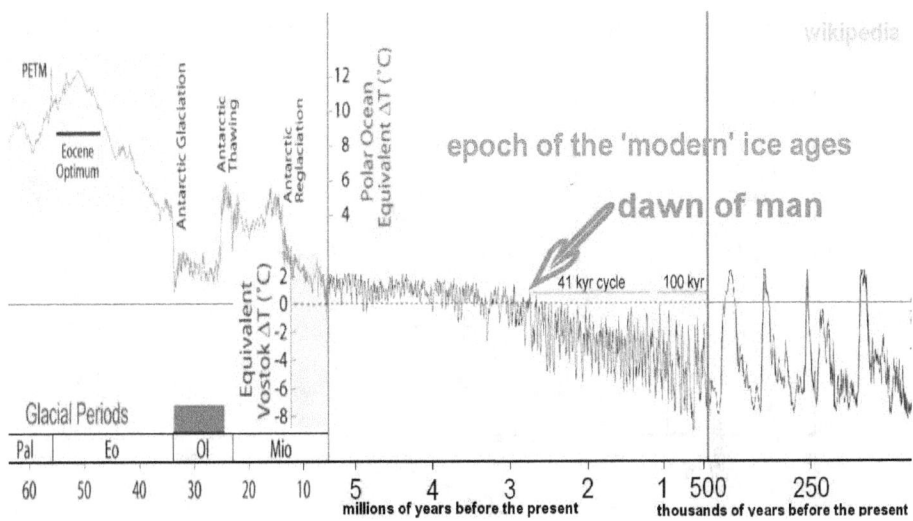

The ice ages didn't last as long in the earlier phase of the weakening conditions. They lasted only 41,000 years then. As the general weakening continued, the 100,000 years long ice ages began. This became named the Pleistocene Epoch.

Throughout the ice ages, in which the Primer Fields fail, the Sun becomes inactive for long periods. It becomes a dim yellow star that glows mostly by its stored up energy and whatever nuclear decay may be ongoing within it.

The Sun remains not totally shut down

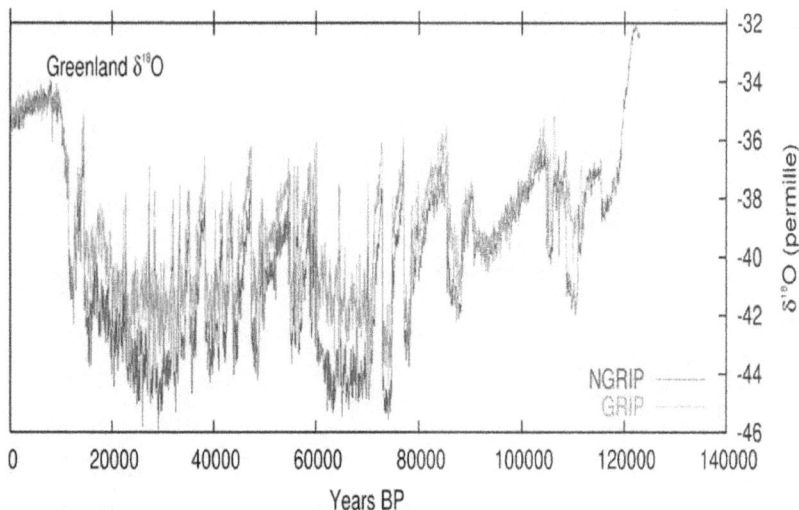

However, evidence exists that the Sun remains not totally shut down during the long ice age glaciation periods, which typically last 100,000 years. The Earth would turn into a snowball if the Sun would remain inactive for 100,000 years. Fortunately, this doesn't happen. Periodically short pulses of high-density conditions do occur during the ice ages, which re-invigorate the plasma streams that enable the Primer Fields to form anew, whereby the Sun to becomes powered again. Unfortunately, these pulses are short in duration. The Sun remains powered by these pulses for only a few decades, and then turns off again.

The Dansgaard-Oeschger oscillations

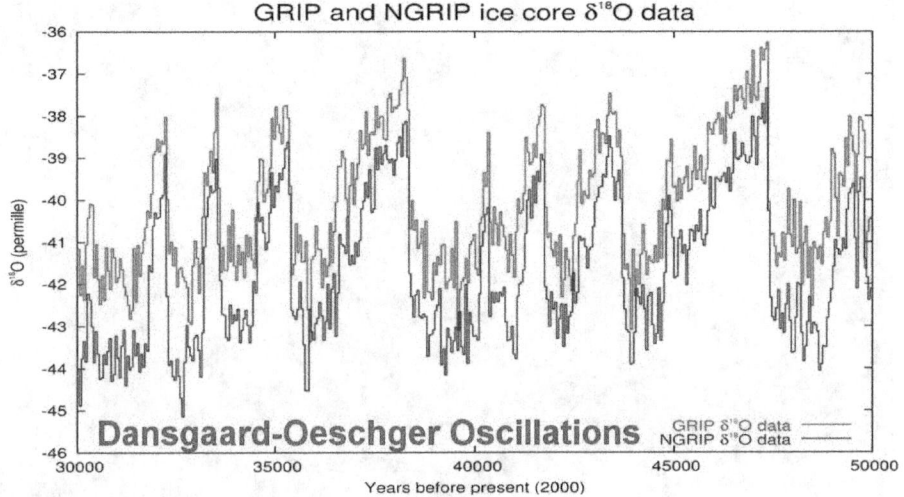

Evidence exists that these pulses occurred on a fairly regular basis. Their occurrence has created large climate oscillations. Evidence has been detected in ice core samples drilled from the Greenland ice sheets that these short periods when the Sun becomes active again, have occurred in intervals of 1470 years. The resulting oscillations have been named, the Dansgaard-Oeschger oscillations.

Part 6

The UFO phenomenon

Transcripts at: www.ice-age-ahead-iaa.ca

The UFO phenomenon

Rapid on-off conditions are natural occurrences

The Dansgaard_Oeschger oscillations are only one type of example in which rapid on-off conditions are natural occurrences for electric plasma phenomena. The so-called UFO phenomenon is another example of the same type.

Many cases of UFO 'sightings' have been reported. Many have been photographed. But note, what has been photographed conforms perfectly to the shape of the electromagnetic bowl-type structure that forms the Primer Fields, which in turn form a sphere of concentrated plasma at their focal point. This happens to be the typical shape that has been captured in UFO photographs.

We see the bowl-type electromagnetic structure, and the plasma sphere at the focal point of it. Both are clearly evident in this photograph. Reports say that the UFOs suddenly appear, that they can stand still, and that they can move quickly. They are seen existing in space, and are even hovering above the moon. They can accelerate immensely fast in their movement, and make extremely sharp sudden turns. They can do all this, because the so-called UFOs are not space ships, but are electromagnetic phenomena that have no mass that would impede such movements. They are merely shapes created by electromagnetic interaction of plasma flowing in the atmosphere, and in space, which can become visible when the conditions are right. Some are seen as moving points of light, even pulsating light, some even appearing in groups.

By being electric plasma concentrations, the UFO phenomenon also reflects radio waves, whereby it can be seen on radar. However, when aircraft are launched to intercept, the electric field around the aircraft affects the field that forms the UFO, which thereby changes its position, or causes it to simply vanish. All this makes interception quite impossible.

No UFO craft does actually exist

NASA - Constellation: trans-lunar injection

Of course interception is also impossible for the simple reason that no UFO craft does actually exist. Extraterrestrial visitors travelling in a physical spacecraft to our planet would have to fly against all the known laws of the physical universe and time. It would take a spacecraft 12,000 years to reach the earth from our closest star, flying at a speed of 360,000 km per hour, which is roughly 30 times the speed of the Apollo 11 flight to the moon, which might be the limit for spacecraft velocities, moving against the plasma background in space. Exotic theories have been invented in attempts to side-step the physical limits and barriers, in attempts to support the UFO theory, while the obvious reality is simply being ignored.

The UFO as an example

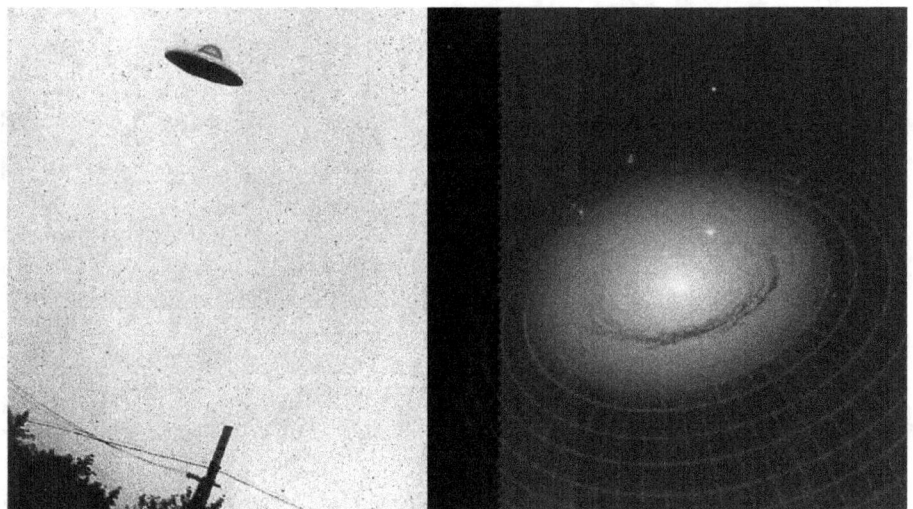

In real terms the UFO phenomenon stands as an example of electromagnetic structures that form in space with gigantic effects and in the atmosphere with minute effects, all formed by Primer Fields that are dynamically created in electric plasma streams.

UFOs comparable to the sprites

The UFOs, thus, become comparable to the sprites that appear in the upper atmosphere. The resulting effects, of course, depend on the coincidence of conditions that enable them, which are inherently rare and fragile in nature, and are specific for the type of atmosphere in which they occur. When the conditions are satisfied in the lower atmosphere, they often enable a number of small plasma events simultaneously, which then become regarded as multiple UFO sightings.

Fragility of the sprites in the stratosphere, and UFO events on the lower atmosphere

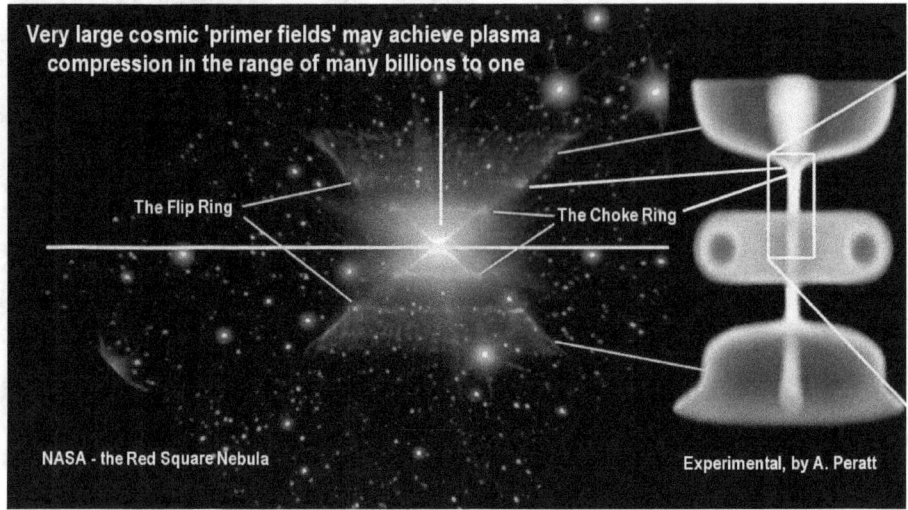

Very large cosmic 'primer fields' may achieve plasma compression in the range of many billions to one

The Flip Ring

The Choke Ring

NASA - the Red Square Nebula

Experimental, by A. Peratt

This fragility of the sprites in the stratosphere, and UFO events on the lower atmosphere evidently also applies to the complex structures and magnetic fields that create the conditions for our Sun to be electrically powered. When the conditions are right, the fields form, and when the conditions no longer exist, the formed fields vanish.

UFO phenomenon fragile

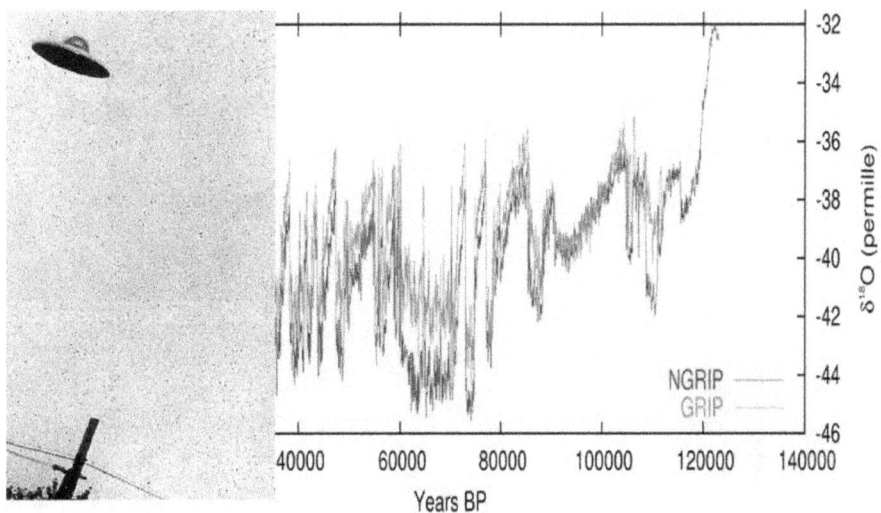

Both the UFO phenomenon and the Dansgaard_Oeschger oscillations illustrate to some degree how fragile the conditions inherently are that affect the powered state of a star, especially that of a small star as our sun.

Powered state of the Sun rare

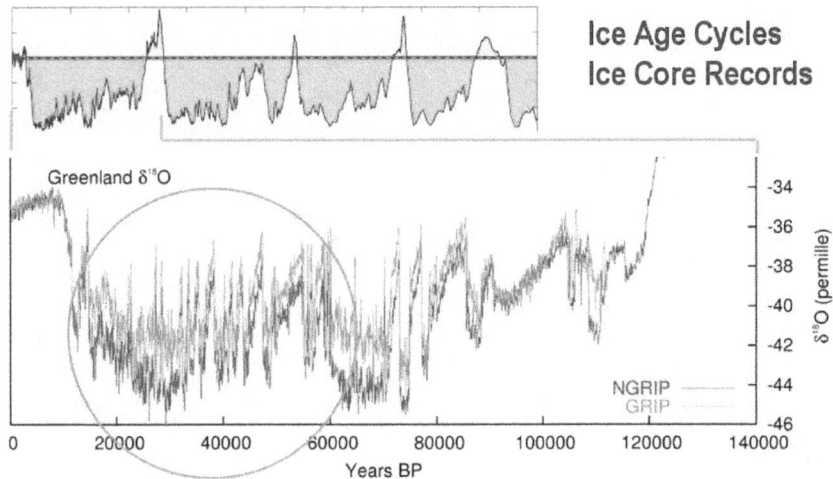

The powered state of the Sun has actually been a rather rare occurrence in the last two million years of the great ice ages. The few long periods that have the Sun constantly active, called the interglacial periods, merely interrupt the 'normal' dark and cold world of the long ice ages that only few people have come through alive.

UFO sightings from the 1950s on

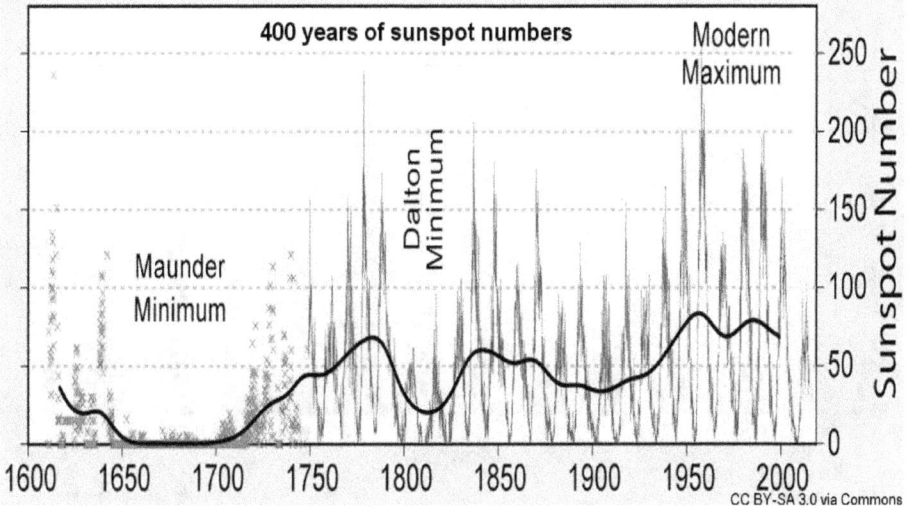

The same fragile nature that applies to our solar system, also applies to the UFO phenomenon. It is not coincidental, therefore, that the UFO sightings became most frequent from the 1950s on, through to the 1990s, when the current Dansgaard_Oeschger pulse peaked, when the solar activity was the strongest in recent time, in which the electric conditions on the Earth were likewise the strongest.

The solar system is on a path to the vanishing point

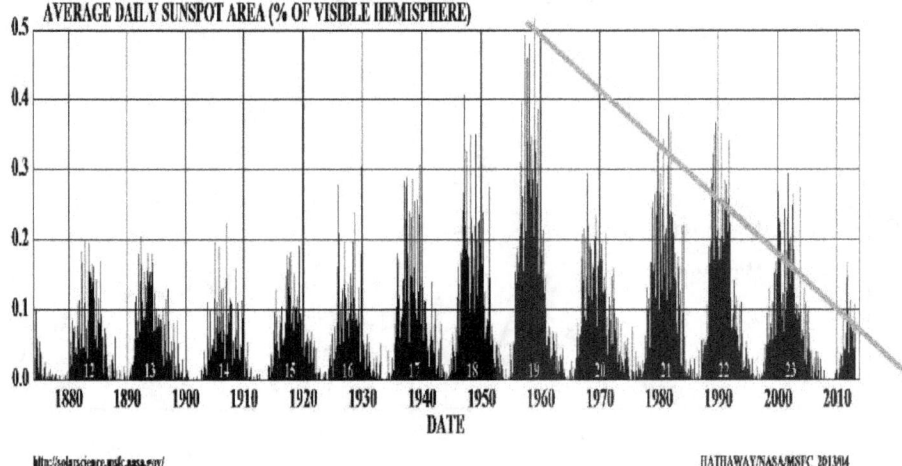

AVERAGE DAILY SUNSPOT AREA (% OF VISIBLE HEMISPHERE)

http://solarscience.msfc.nasa.gov/

HATHAWAY/NASA/MSFC 2013/04

With the electric environment around our Sun now getting rapidly weaker again, according to the down-ramping that the Ulysses satellite has reported, we appear to have entered a transition zone of extremely fragile conditions. The solar system, as an UFO, is on a path to the vanishing point.

The critical warning

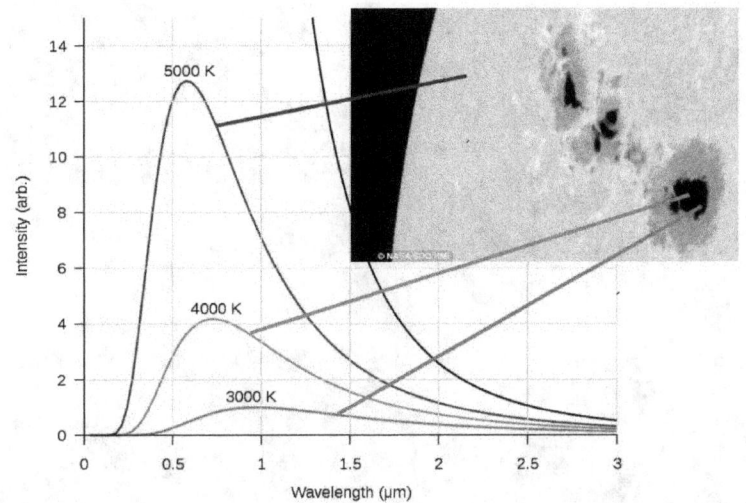

The critical warning that science affords us on this basis, unfortunately, is not precise to the day, nor in exact details. Still, the down-ramping that one sees in many areas happening simultaneously presents strong points of scientific correlation of the numerous elements that the astrophysical arena can provide, that points towards a big phase shift in the making.

The exact day for the Sun reverting to its unpowered state, cannot be determined. But do we need to know the day of the event when the principle is known and the dynamics of its manifestation? The evidence that we have before us, presents a strong case for us all to make extreme efforts towards building the vast new infrastructures that our continued existence on this planet WILL most certainly depend on.

The entire world is affected

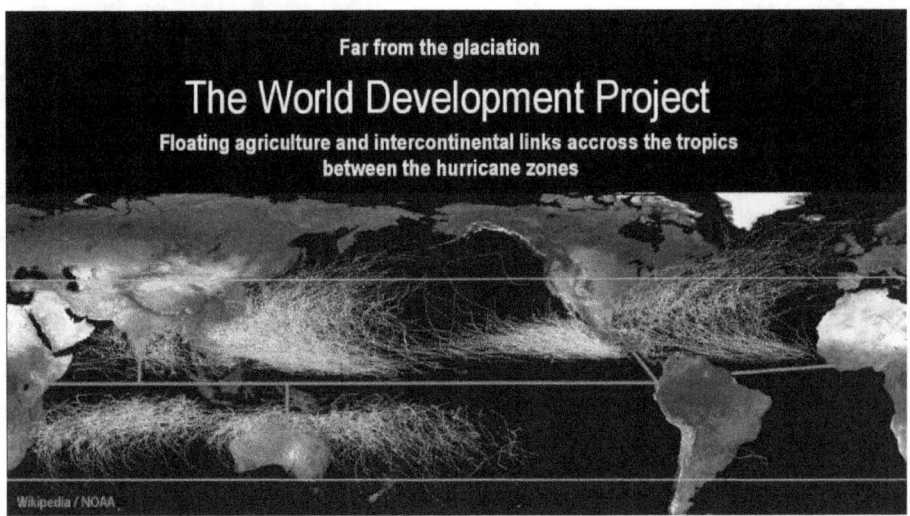

It is tempting to assume that only the far northern countries and
regions, like Canada, Europe, Russia, and China, are affected by the
coming ice age transition in which the Sun becomes inactive. This is
a foolish assumption. When the Sun turns dim and the food supply
is collapsing, the entire world is affected. Thus it becomes the task
of humanity as a whole to build the infrastructures to protect
human existence on this planet. Not to act decisively on this front
amounts to committing universal suicide. The changing
astrophysical environment that affects our world will affect
everything. It affects even the magnitude of earthquakes. It is highly
likely on this front too, that we haven't seen anything yet in terms
of consequences, which of course we can avoid by placing our
civilization afloat onto the sea.